Advanced MR Neuroimaging

Series in Medical Physics and Biomedical Engineering

Series Editors: John G. Webster, E. Russell Ritenour, Slavik Tabakov, and Kwan-Hoong Ng

Advanced MR Neuroimaging
from Theory to Clinical Practice

Ioannis Tsougos
Assistant Professor of Medical Physics
Faculty of Medicine, School of Health Sciences,
University of Thessaly
Biopolis, Larissa, Greece

CRC Press
Taylor & Francis Group
Boca Raton London New York

CRC Press is an imprint of the
Taylor & Francis Group, an **informa** business

CRC Press
Taylor & Francis Group
6000 Broken Sound Parkway NW, Suite 300
Boca Raton, FL 33487-2742

First issued in paperback 2019

ISBN-13: 978-1-4987-5523-8 (hbk)
ISBN-13: 978-0-367-87650-0 (pbk)

Library of Congress Cataloging-in-Publication Data

Names: Tsougos, Ioannis, author.
Title: Advanced MR neuroimaging : from theory to clinical practice / Ioannis Tsougos.
Other titles: Series in medical physics and biomedical engineering.
Description: Boca Raton, FL : CRC Press, Taylor & Francis Group, [2018] |
Series: Series in medical physics and biomedical engineering | Includes bibliographical references and index.
Identifiers: LCCN 2017037811| ISBN 9781498755238 (hardback ; alk. paper) |
ISBN 1498755232 (hardback ; alk. paper) | ISBN 9781498755252 (e-book) |
ISBN 1498755259 (e-book)
Subjects: LCSH: Brain–Magnetic resonance imaging. | Magnetic resonance imaging.
Classification: LCC RC386.6.M34 T73 2018 | DDC 616.8/04754–dc23
LC record available at https://lccn.loc.gov/2017037811

Visit the Taylor & Francis Web site at
http://www.taylorandfrancis.com

and the CRC Press Web site at
http://www.crcpress.com

To Mary, Achilles-Meletis and Eleni

Contents

7 Functional Magnetic Resonance Imaging (fMRI)

8 Artifacts and Pitfalls of fMRI

9 The Role of Multiparametric MR Imaging—Advanced MR Techniques in the Assessment of Cerebral Tumors

Series Preface

The *Series in Medical Physics and Biomedical Engineering* describes the applications of physical sciences, engineering, and mathematics in medicine and clinical research.

The series seeks (but is not restricted to) publications in the following topics:

- Artificial organs
- Assistive technology
- Bioinformatics
- Bioinstrumentation
- Biomaterials
- Biomechanics
- Biomedical engineering
- Clinical engineering
- Imaging
- Implants
- Medical computing and mathematics

- Medical/surgical devices
- Patient monitoring
- Physiological measurement
- Prosthetics
- Radiation protection, health
- physics, and dosimetry
- Regulatory issues
- Rehabilitation engineering
- Sports medicine
- Systems physiology
- Telemedicine
- Tissue engineering
- Treatment

The *Series in Medical Physics and Biomedical Engineering* is an international series that meets the need for up-to-date texts in this rapidly developing field. Books in the series range in level from introductory graduate textbooks and practical handbooks to more advanced expositions of current research.

The *Series in Medical Physics and Biomedical Engineering* is the official book series of the International Organization for Medical Physics.

The International Organization for Medical Physics

The International Organization for Medical Physics (IOMP) represents over 18,000 medical physicists worldwide and has a membership of 80 national and 6 regional organizations, together with a number of corporate members. Individual medical physicists of all national member organisations are also automatically members.

The mission of IOMP is to advance medical physics practice worldwide by disseminating scientific and technical information, fostering the educational and professional development of medical physics and promoting the highest quality medical physics services for patients.

A World Congress on Medical Physics and Biomedical Engineering is held every three years in cooperation with International Federation for Medical and Biological Engineering (IFMBE) and International Union for Physics and Engineering Sciences in Medicine (IUPESM). A regionally based international conference, the International Congress of Medical Physics (ICMP) is held between world congresses. IOMP also sponsors international conferences, workshops and courses.

The IOMP has several programmes to assist medical physicists in developing countries. The joint IOMP Library Programme supports 75 active libraries in 43 developing countries, and the Used Equipment Programme coordinates equipment donations. The Travel Assistance Programme provides a limited number of grants to enable physicists to attend the world congresses.

IOMP co-sponsors the *Journal of Applied Clinical Medical Physics*. The IOMP publishes, twice a year, an electronic bulletin, *Medical Physics World*. IOMP also publishes e-Zine, an electronic news letter about six times a year. IOMP has an agreement with Taylor & Francis for the publication of the *Medical Physics and Biomedical Engineering* series of textbooks. IOMP members receive a discount.

IOMP collaborates with international organizations, such as the World Health Organisations (WHO), the International Atomic Energy Agency (IAEA) and other international professional bodies such as the International Radiation Protection Association (IRPA) and the International Commission on Radiological Protection (ICRP), to promote the development of medical physics and the safe use of radiation and medical devices.

Guidance on education, training and professional development of medical physicists is issued by IOMP, which is collaborating with other professional organizations in development of a professional certification system for medical physicists that can be implemented on a global basis.

The IOMP website (www.iomp.org) contains information on all the activities of the IOMP, policy statements 1 and 2 and the 'IOMP: Review and Way Forward' which outlines all the activities of IOMP and plans for the future.

Preface

Since its early medical application about 40 years ago, magnetic resonance imaging has revolutionized brain neuroimaging, providing non-invasively excellent high-resolution images without the use of ionizing radiation.

Nevertheless, despite the superior quality, conventional MR imaging provides only anatomical, rather than physiological, information and may therefore be sometimes non-specific.

During the last decade, some of the greatest achievements in neuroimaging have been related to remarkable advances in MR techniques, which provided insights into tissue microstructure, microvasculature, metabolism, and brain connectivity. These advanced MR neuroimaging techniques include diffusion, perfusion, magnetic resonance spectroscopy, and functional MRI.

Previously available mostly in research environments, they are now establishing themselves firmly in the everyday clinical practice in a plethora of clinical MR systems. However, despite the growing interest and wider acceptance, the lack of a comprehensive body of knowledge, the intrinsic complexity and physical difficulty of the techniques, as well as an appreciable number of associated artifacts and pitfalls, still confine their routine clinical application.

This book focuses on the basic principles and physics theory of diffusion, perfusion, magnetic resonance spectroscopy, and functional MRI, accompanied by their clinical applications, with particular emphasis on the associated artifacts and pitfalls using a comprehensive and didactic approach. It aims to bridge the gap between theoretical applications and optimized clinical practice of advanced techniques by addressing all of them in a single and concise volume.

The book is organized in nine chapters. Four chapters (Chapters 1, 3, 5, and 7) describe the basic principles of the discussed techniques, providing an overview of the methods with a step-by-step didactic approach, explaining fundamentals as well as clinical implications. These are, followed by respective dedicated chapters on the potential artifacts and pitfalls of each technique, including the proposed mitigating strategies, with special attention in the post-processing techniques (Chapters 2, 4, 6, and 8).

The final chapter (Chapter 9) covers a multiparametric approach utilizing all the aforementioned advanced MR techniques, evaluating the different underlying patho-physiological characteristics of brain tumors in an attempt to illustrate the potential ability of these techniques to contribute to a more accurate diagnosis.

Each chapter, as well as several important sections within each chapter, begins with a dedicated "Focus Point" box, "sensitizing" the reader's attention to the key features, by highlighting the most important concepts that follow. An introduction in every chapter further guides the reader into the forthcoming more detailed information. Lastly, summary tables and aggregated classifications are used in an attempt to facilitate memorization, supporting those who wish to delve into and apply these techniques in the clinical routine.

The book can serve as an educational manual for neuroimaging researchers and basic scientists (radiologists, neurologists, neurosurgeons, medical physicists, engineers, etc.) with an interest in advanced MR techniques, as well as a reference for experienced clinical scientists who wish to optimize their multi-parametric imaging approach.

In conclusion, I sincerely hope this is an easy-to-read yet comprehensive handbook, which can be used as an essential guide to the advanced MR imaging techniques routinely used in clinical practice for the diagnosis and follow-up of patients with brain tumours.

Ioannis Tsougos
Assistant Professor of Medical Physics
Medical School, University of Thessaly

About the Author

Dr. Ioannis Tsougos holds a BSc in physics, and an MSc and a PhD in medical radiation physics. Currently, he is an assistant professor of Medical Radiation Physics at the medical school of the University of Thessaly, Larissa, Greece and a visiting researcher in the Neuroimaging Division at the Institute of Psychiatry, Psychology, and Neuroscience, King's College London, London, United Kingdom. He has authored more than 75 research papers and 10 international book chapters. In addition, he frequently acts as a reviewer for several journals in medical physics/radiology and European research foundations projects. Dr. Tsougos has broad multidisciplinary teaching and clinical experience, specializing in advanced MR techniques, and he is a member of the EFOMP, ESR, and ESMRMB.

Diffusion MR Imaging

1.1 Introduction

1.1.1 Diffusion

Focus Point

- Particles suspended in a fluid (liquid or gas) are forced to move in a random motion called "Brownian motion."
- Diffusion is "Brownian motion."

Diffusion refers to the random, microscopic movement of particles due to thermal collisions. Particles suspended in a fluid (liquid or gas) are forced to move in a random motion, which is often called "Brownian motion" or *pedesis* (from Greek: πήδησις [meaning "leaping"]) resulting from their collision with the atoms or molecules in the gas or liquid.

This diffuse motion was named after Robert Brown, the famous English botanist, who observed under a microscope that pollen grains in water were in a constant state of agitation. It was as early as 1827 and, unfortunately, he was never able to fully explain the mechanisms that caused this motion. He initially assumed that he was observing something "alive," but later he realized that something else was the cause of this motion since he had detected the same fluctuations when studying dead matter such as dust.

Atoms and molecules had long been theorized as the constituents of matter, and many decades later (in 1905) Albert Einstein published a paper explaining in precise detail how the motion that Brown had observed was a result of the pollen being moved by individual water molecules (Einstein, 1905). In the introduction of his paper, it is stated that

> ... according to the molecular-kinetic theory of heat, bodies of a microscopically visible size suspended in liquids must, as a result of thermal molecular motions, perform motions of such magnitudes that they can be easily observed with a microscope. It is possible that the motions to be discussed here are identical with so-called Brownian molecular motion; however, the data available to me on the latter are so imprecise that I could not form a judgment on the question ....

To get a feeling of the physical meaning of diffusion, consider a diffusing particle that is subjected to a variety of collisions that we can consider random, in the sense that each such event is virtually unrelated to its previous event. It makes no difference whether the particle is a molecule of perfume diffusing in air, a solute molecule in a solution, or a water molecule inside a medium diffusing due to the medium's thermal energy.

Einstein described the mathematics behind Brownian motion and presented it as a way to indirectly confirm the existence of atoms and molecules in the formulation of a diffusion equation, in which the diffusion coefficient is related to the mean squared displacement of a Brownian particle.

In other words, Einstein sought to determine how far a Brownian particle travels in a given time interval.

For this purpose, he introduced the "displacement distribution," which quantifies the fraction of particles that will traverse a certain distance within a particular timeframe, or equivalently, the likelihood that a single given particle will undergo that displacement.

Using this concept, Einstein was able to derive an explicit relationship between displacement and diffusion time in the following equation:

$$\langle x^2 \rangle = 6Dt$$

(1.1)

where $\langle x^2 \rangle$ is the mean-squared displacement of particles during a diffusion time t, and D is the diffusion coefficient. The distribution of squared displacements takes a Gaussian form, with the peak being at zero displacement and with equal probability of displacing a given distance from the origin no matter in which direction it is measured. Actually, the Gaussian diffusion can be calculated in one, two, or three dimensions. The form of the Gaussian in one dimension is the familiar bell-shaped curve and the displacement is $2Dt$. In two dimensions, if the medium is isotropic, the cross-section of the curve is circular, with the radius given by $4Dt$, centered on the origin. When extended to three dimensions, the iso-probability surface is a sphere, of radius $6Dt$ as in Equation 1.1, and again centered on the origin.

The concept of diffusion can be easily demonstrated by adding a few drops of ink to a glass of water. The only pre-requirement is for the water in the glass to be still. Initially, the ink will be concentrated in a very small volume, and then with time, it will diffuse into the rest of the water until the concentration of the ink is uniform throughout the glass. The speed of this process of diffusion, or the rate of change of concentration of the ink, gives a measure of the property of medium where diffusion takes place. In that sense, if we could follow the diffusion of water molecules into the brain, we would reveal aspects of functionality of the normal brain tissue itself. More importantly, by understanding in more detail normal brain functionality, we would then be able to analyze the kind of changes that may occur in the brain when it is affected by various disease processes.

In other words, diffusion properties represent the microscopic motion of water molecules of the tissue; hence it can be used to probe local microstructure. As water molecules are agitated by thermal energy, they diffuse inside the body, hindered by the boundaries of the surrounding tissues or other biological barriers. By probing this movement, the reconstruction of the boundaries that hinder this motion can be visualized.

1.1.2 Diffusion in Magnetic Resonance Imaging

Magnetic resonance imaging (MRI) with its excellent soft tissue visualization and variety of imaging sequences has evolved to one of the most important noninvasive diagnostic tools for the detection and evaluation of the treatment response of cerebral tumors. Nevertheless, conventional MRI presents limitations regarding certain tumor properties, such as infiltration and grading (Hakyemez et al., 2010). It is evident that a more accurate detection of infiltrating cells beyond the tumoral margin and a more precise tumor grading would strongly enhance the efficiency of differential diagnosis Diffusion-weighted imaging (DWI) provides noninvasively significant structural information at a cellular level, highlighting aspects of the underlying brain pathophysiology.

In theory, DWI is based on the freedom of motion of water molecules, which can reflect tissue microstructure; hence the possibility to characterize tumoral and peritumoral microarchitecture, based on water diffusion findings, may provide clinicians a whole new perspective on improving the management of brain tumors. Although, initially, DWI was established as an important method in the assessment of stroke (Schellinger et al., 2001), a large number of studies have been conducted in order to assess whether the quantitative information derived by DWI may aid differential diagnosis and tumor grading (Fan et al., 2006; Lam et al., 2002; Kono et al., 2001; Yamasaki et al., 2005), especially in cases of ambiguous cerebral neoplasms (Nagar et al., 2008). Moreover, DWI may also have a significant role in therapeutic follow-up and prognosis establishment in various brain lesions. Given its important clinical role, DWI should be an integral part of diagnostic brain imaging protocols (Schmainda, 2012; Zakaria et al., 2014).

1.2 Diffusion Imaging: Basic Principles

1.2.1 Diffusion-Weighted Imaging

Focus Point

- Particles suspended in a fluid (liquid or gas) are forced to move in a random motion, which is often called "Brownian motion."
- Diffusion is considered the result of the random motion of water molecules.
- Molecular diffusion in tissues is not free, but reflects interactions with many obstacles, such as macromolecules, fibers, membranes, etc.
- By understanding normal brain diffusion, we would be able to analyze the kind of changes that may occur in the brain when it is affected by various disease processes.
- DWI represents the microscopic motion of water molecules hence probes local tissue microstructure.

As already explained, diffusion is considered the result of the random walk of water molecules inside a medium due to their thermal energy, and is described by the "Brownian" law by a diffusion constant D. Water makes up 60%–80% of human body weight. For pure water at ~37°C, D is approximately $3.4 \times 10{-3}$ mm^2/s (Gillard et al., 2005). In an isotropic medium, diffusion is equally distributed towards all directions, described previously as the drop of ink in a glass of water. Nevertheless, it is evident that within an anisotropic medium such as human tissue, water motion will be restricted. Therefore, inside an even more complex environment, such as the human brain, cell membranes, neuronal axons, and other macromolecules act as biological barriers to free water motion, hence water mobility is considered anisotropic. In other words, in the brain, water molecules bounce, cross, and interact with tissue components. Therefore, in the presence of those obstacles, the actual diffusion distance is reduced compared to free water, and the displacement distribution is no longer Gaussian. Strictly speaking, while over very short times, diffusion reflects the local intrinsic viscosity, at longer diffusion times the effects of the obstacles become predominant. Although the observation of this displacement distribution is made on a statistical basis, it provides unique clues about the structural features and geometric organization of neural tissues on a microscopic scale, as well as changes in those features with physiological or pathological states (Le Bihan et al., 2006).

More specifically, the highly organized white-matter bundles, due to their myelin sheaths, force water to move along their axes, rather than perpendicular to them, as an apt analogy of a bundle of cables as seen in Figure 1.1.

Hence, MR can be used to probe the structural environment providing a unique opportunity to visually quantify the diffusional characteristics of tissue. This is of paramount importance in the environment of a biological sample in which the size of the area under study is so small that conventional imaging techniques are insufficient. Try to keep in mind that in 50 ms (this is considered a "typical" time interval for diffusion measurements) the diffusion distance of "free" water molecules at 37°C, will be about **17 μm**. DWI is an advanced MR technique, which is based on the aforementioned Brownian motion of molecules to acquire images. One must not forget, however, that the overall signal observed in a "diffusion" MR image volume element (voxel), at a millimetric resolution, results from the integration, on a statistical basis, of all the microscopic displacement distributions of the water molecules present in this voxel. With most current MRI systems, especially those developed for human applications, the voxel size remains quite large (that is, a few mm³). The averaging and smoothing effect resulting from this scaling presumes some homogeneity in the voxel and makes it difficult to obtain a direct physical interpretation from the global parameter, unless some assumptions can be made (Le Bihan et al., 1986, 2006). The exact relationship between the diffusion properties and specific tissue microscopic features is currently the object of intensive research (Kaden et al., 2016).

> Let us try and see how diffusion affects the MR signal and how this can be measured and evaluated in clinical practice.

When a patient is inserted into the homogeneous magnetic field of an MR scanner, the nuclear spins are lined up along the direction of the static magnetic field. Nevertheless, there is no such thing as a perfectly homogeneous magnetic field because it simply can't be produced.

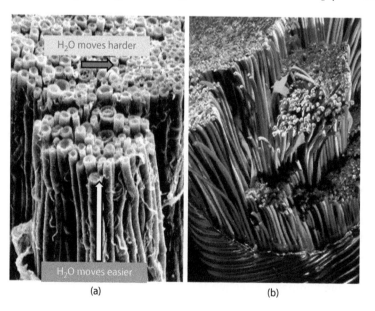

(a)   (b)

FIGURE 1.1   Freeze-fractured section through a bundle of myelinated nerve fibers (a) and the apt analogy of a bundle of cables (b). It is evident that water molecules will move more easily along the direction of the nerve fibers, rather than perpendicular to them.

Even if it could, the insertion of the patient's susceptibility effects (such as the sinuses or bone, etc.) would make it inhomogeneous.

The effect of these external field inhomogeneities on the self-diffusion of molecules was first reported by Erwin Hahn (Hahn, 1950) who observed that: "... *nuclear signals due to precessing nuclear moments contained in liquid molecules are not only attenuated by the influence of T1 and T2, but also suffer a decay due to the self-diffusion of the molecules into differing local fields established by external field inhomogeneities.*"

At the same time, Bloch (Bloch, 1950) realized that the effects of diffusion can be magnified by purposely imposing a field inhomogeneity in a controlled manner. In that sense, if a radio-frequency pulse is applied, the protons will spin at different rates depending on the strength, duration, and direction of the so-called "gradient." By applying an equal and opposite gradient, the protons will be refocused, hence information about how much the nuclear spins have moved (diffused) during this time can be acquired. In other words, stationary protons will provide a null signal after this counter-process while mobile protons, which have changed position between the two gradients, will present a signal loss.

Moving forward, in DW MRI, we simply measure the dephasing of proton spins in the presence of a gradient field (i.e., a magnetic field that spatially varies). Hence, the basic phenomenon studied is the change of a proton's phases along the axis of the applied gradient field, which is expected to increase with "diffusion time." The longer and stronger the gradient pulses, the more direction changes of the molecules and hence the bigger the loss of coherence and signal attenuation because of the macroscopic motion. By comparing the signal amplitude with and without the diffusion-encoding gradient applied, the portion of dephasing resulting from incoherent motion during the application of the gradient can be isolated (Jones et al., 2013).

A schematic of this sequence is presented in Figure 1.2.

It should be clear now that the brightness of each voxel of a diffusion image corresponds to the DW intensity, which in turn corresponds to the amount of diffusion weighting or gradient.

Hence, tissues closer to water (e.g., CSF) that have mobile protons, would give lower intensity while more static or solid tissues (e.g., white matter) would give a stronger signal. In that sense, DW contrast behaves like T1 weighting, or more precisely, like inverse T2 weighting. Figure 1.3 depicts an axial T2 image (a), with the corresponding T1 image (b), DW image (c), and ADC image (d), collected from a male subject.

1.2.2 The b-Value

The aforementioned signal loss would be dependent on the degree of diffusion weighting, which is referred to as the b-value. The b-value is a factor that reflects the strength and duration of the gradient pulses used to generate DW images. Therefore, to put it as simply as possible, ignoring the stationary protons and measuring the signal of the mobile protons, the amount of diffusion that has occurred in a specific direction can be determined. The b-value is described by the following mathematical equation:

$$b = (\gamma G \delta)^2 \left(\Delta - \frac{\delta}{3} \right) \tag{1.2}$$

A more detailed explanation of the b-factor is obviously out of the "clinical" scope of this book, but this expression can be used to obtain an estimate of the diffusion "sensitization" for a given experiment. In this equation, Δ is the temporal separation of the gradient pulses, δ is their duration, G is the gradient amplitude, and γ is the gyromagnetic ratio of protons (= 42.58 MHz/T) (Stejskal and Tanner, 1965). The diffusion time is assigned as ($\Delta - \delta/3$), where the second term in the expression accounts for the finite duration of the pulsed field gradients. The units

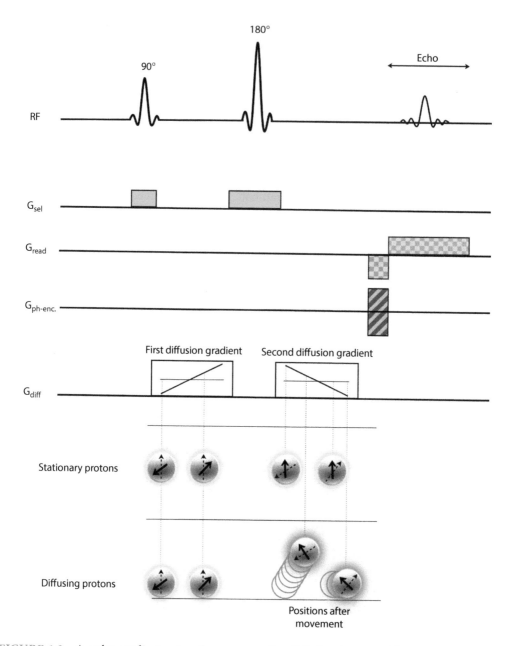

FIGURE 1.2 A pulse-gradient spin-echo sequence for diffusion imaging. The standard spin-echo sequence is diffusion sensitized using a gradient pulse pair (Gdiff) so that the spin phase-shift depends on location along the first gradient pulse. The 180° radiofrequency (RF) pulse and the second gradient pulse will rephase the static spins while diffusing spins will be "caught" out of phase.

for the b-value are s/mm^{-2}, and the range of values typically used in clinical diffusion weighting is 800–1500 s/mm^{-2}. The formula for the b-factor implies that we can increase diffusion weighting (DW), or "sensitization," by increasing either gradient timing, δ or Δ, or gradient strength, G. Note that the equation for the b value does not take into account the rising and falling edges of the diffusion gradients (see Figure 1.2) and assumes perfect rectangles. This is not actually the case, but we will discuss that in the next chapter.

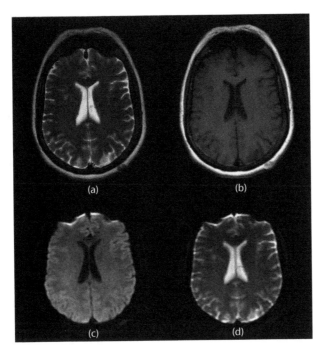

FIGURE 1.3   Axial T2 Image (a), T1 Image (b), DW image (c), and ADC image (d) collected from a male subject.

It can be shown that for a fixed diffusion weighting, the signal in a DW experiment is given by the following equation:

$$S = S_0 e^{\frac{TE}{T2}} e^{-bD} \quad \text{or} \quad S = S_0 e^{-bADC} \tag{1.3}$$

So inevitably, at this point, the question arises: What is the optimal b-value for clinical DWI?

It has been shown that for typical imaging experiments, the optimal b-value is about 1257 s/mm^{-2} (Jones et al., 1999). However, most studies are limited to DW imaging using b-factors of 0 and 1000. An upper b-factor around 1000 s/mm^{-2} has been available for most clinical scanners until now and DW imaging using these standard values has been shown to be effective in detecting and delineating restricted diffusion, for example, in acute ischemic lesions of the brain. Nevertheless, it may be more important to consider that in a clinical setting, it is advisable to maintain the same b-value for all examinations, making it easier to learn to interpret these images and become aware of the appearance of findings in various disease processes.

However, in more recent studies, it has been proposed that high b-value diffusion MR imaging can provide enhanced contrast toward different cellular components when compared to 1000 s/mm^2. It is argued that the parallel analysis of low and high b-value diffusion images may provide a more comprehensive tissue characterization, enabling improved sensitivity and specificity of DW MRI to healthy tissue microstructure and subtle pathology, especially in white matter (Assaf and Basser, 2005; Ben-Amitay et al., 2012; Seo et al., 2008).

Unfortunately, such analyses may suffer from poor signal to noise ratio, leading to longer acquisition times and therefore more motion artifacts, limiting their clinical application. For more details please refer to Chapter 2.

1.2.3 Apparent Diffusion Coefficient

In Equation 1.3, S_0 is the signal intensity in the absence of any T2 or diffusion weighting, *TE* is the echo time and *D* is the apparent diffusivity, usually called the Apparent Diffusion Coefficient (ADC). The term "apparent" reveals that, because tissues have a complicated structure, it is often an average measure of a number of multiple incoherent motion processes and does not necessarily reflect the magnitude of intrinsic self-diffusivity of water (Le Bihan et al., 1986; Tanner, 1978).

Hence, to reflect the fact that we are not talking about the intrinsic self-diffusivity of water, and to clarify that this estimated diffusivity comes from a sum of different spins, we use the term ADC. Please note that ADC is a calculated value based on diffusion images using at least two different b-values as shown by the following equation:

$$\text{ADC image} = \left(-\frac{1}{b}\right)\ln\left(\frac{DW\,\text{image}}{T2_w\,\text{image}}\right) \tag{1.4}$$

For a more accurate value of ADC, instead of two, a range of b-values can be applied and a "least-squares" fit can be performed. ADC can also be displayed as a colored map as seen in Figure 1.4.

Going back to the b-value, the first exponential term in Equation 1.2 is the weighting due to transverse (T2) relaxation and the second term shows that diffusion induces an exponential attenuation to the signal (Price, 2007). As the diffusing spins are moving inside the field gradient, each spin is affected differently by the field, thus the alignment of the spins with each other is altered. Since the measured signal is a summation of tiny signals from all individual spins, the

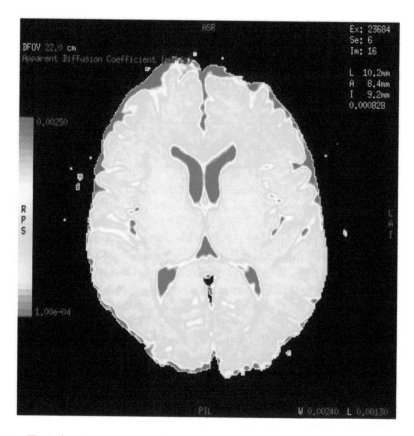

FIGURE 1.4   Typical ADC parametric color map of a healthy volunteer.

misalignment, or "dephasing," caused by the gradient pulses results in a drop in signal intensity; the longer the diffusion distance, the more dephasing, the lower the signal (Moritani et al., 2009). The goal of DWI is to estimate the magnitude of diffusion within each voxel, i.e., the tissue micro-structure, and this can be measured by the term ADC. The aforementioned parametric map of ADC values is obtained in order to facilitate qualitative measurements. The intensity of each image pixel on the ADC map reflects the strength of diffusion in the pixel. Therefore, a low value of ADC (dark signal or "cold" color) indicates restricted water movement, whereas a high value (bright signal or "warm" color) of ADC represents free diffusion in the sampled tissue (Debnam and Schellingerhout, 2011). A "quick" way to remember DWI and ADC is depicted in Figure 1.5.

A high ADC value implies high motion (free diffusion) and therefore low signal in a DW image.

For example, as seen in Figure 1.3, in cerebral regions where water diffuses freely, such as CSF inside the ventricles, there is a drop in signal on the acquired DW images, whereas in areas that contain many more cellular structures and constituents (gray matter or white matter), water motion is relatively restricted and the signal on DW images is increased. Consequently, regions of CSF will present higher ADC values than other brain tissues on the parametric maps. ADC is measured in units of mm²s⁻¹. An indicative value of ADC for pure water at room temperature is approximately 2.2×10^{-3} mm²s⁻¹. Typical normal and pathological tissue ADC values are given in Table 1.1.

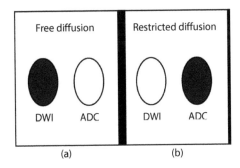

FIGURE 1.5 Signal intensities of DWI and ADC relative to diffusion characteristics. When diffusion is not restricted, the DWI signal is low and ADC signal is high (a). When there is restriction in diffusion the DWI signal is high and ADC is low (b).

TABLE 1.1 ADC Values ($\times 10^{-3}$ mm²/s) in the Normal Brain and Indicative Disorders

ROI	ADC value ($\times 10^{-3}$mm²/s)
Normal Brain	
White matter	0.84 ± 0.11
Corpus callosum	0.75 ± 0.15
CSF	3.40 ± 0.45
Thalamus	0.83 ± 0.14
Pons	0.84 ± 0.15
Cerebellar parenchyma	0.83 ± 0.17
Disorder	
Acute infarct (cytotoxic edema)	0.32 ± 0.09
Vasogenic edema	1.68 ± 0.27
Glioblastoma multiforme	1.27 ± 0.46
Brain metastasis	1.17 ± 0.52

1.2.4 Isotropic or Anisotropic Diffusion?

At this point it is important to clarify the concept of isotropy or anisotropy. Isotropy is derived from the Greek word *isos* (ἴσος, meaning "equal") and *tropos* (τρόπος, meaning "way") thus meaning "equal way" or uniform in all orientations. On the contrary, exceptions, or inequalities in Greek, are frequently indicated by the prefix "an" (meaning "the opposite of"), hence the term "anisotropy." In that sense, anisotropy is used to describe situations where properties vary systematically, dependent on direction. An attempt to visualize a completely isotropic and gradually anisotropic voxel is depicted in Figure 1.6.

In pure water, molecules are equally likely to move in any direction, therefore, water's diffusion properties should be isotropic. This would mean that the MR signal will be absolutely the same, irrespective of the physical direction of the applied gradients. Indeed, this is the case as shown in Figure 1.7, where three different DW images of a water phantom are depicted, one for each of the principal axes of the scanner, X, Y, and Z. But in many biological tissues, diffusion is restricted to certain directions because of the cell membranes and other organelles, for example, in directional structures such as the nerve fibers, where diffusion is preferential along the fibers rather than across them (Figure 1.1).

Areas of the brain with similar diffusion properties in every direction are said to be isotropic and independent of the direction of application of the diffusion gradients; they will have the same signal characteristics on DW images. On the other hand, anisotropic areas are characterized by different diffusion coefficients in different directions; in these cases, the signal attenuation reflects the diffusion properties in the direction of application of the diffusion gradients.

The measurement of the degree of this diffusion isotropy reveals aspects of the tissue's microstructure, for example, the degree of myelination of the nerve fibers. The effect of isotropy or anisotropy is shown in Figure 1.7b, where the DW images of the three principal axes gradients of the scanner are depicted for a healthy volunteer. The DW intensity of certain regions of the brain is the same in all three images, suggesting that the ADC is the same in all directions. Thus, diffusion can be assumed to be isotropic. However, in other regions, for example, the corpus callosum, the diffusion is clearly anisotropic since there are differences among the three different images, representing directionality of the local microstructure.

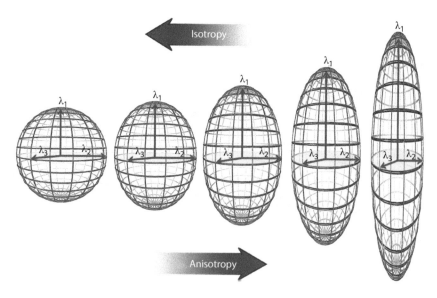

FIGURE 1.6   Attempt to visualize a completely isotropic and gradually anisotropic voxel.

Indeed, the ordered structure of the corpus callosum in a left-right orientation can be seen as a high ADC value in the left image of Figure 1.7 since the diffusion-encoding gradients were applied in the same orientation. On the contrary, in the other two images, the region of the corpus-callosum has lower ADC values, indicating that diffusion is relatively hindered along these directions.

So, what is the source of this diffusion anisotropy?

There were several initial suggestions for the mechanisms mediating diffusion isotropy or anisotropy, including the myelin sheath (Thomsen et al., 1987; Beaulieu and Allen, 1994a, Beaulieu and Allen, 1996), local susceptibility gradients (Hong and Dixon, 1992; Lian et al., 1994), axonal cytoskeleton, and fast-axonal transport. Nevertheless, in a more recent work by Beaulieu (2002), it is reported that the main determinant of anisotropy in nervous tissue is the presence of intact cell membranes and that myelination only serves to modulate anisotropy.

Now, it should be clear that in the case of the destruction of biological barriers such as the cell membranes, ADC should increase as isotropy increases. Hence, it follows that one should expect an increase of ADC in disease as destruction of tissue generally reduces anisotropy. This can be illustrated in Figure 1.8 in a case of a high-grade glioma. The axial T2-FLAIR (a) and T1-weighted post-contrast (b) images demonstrate a right temporal lesion with surrounding edema and ring-shaped enhancement. On the DW image (c), the lesion presents low signal intensity, resulting in high intratumoral ADC (d). The relatively high ADC of the peritumoral edema reflects tumor infiltration in the surrounding parenchyma.

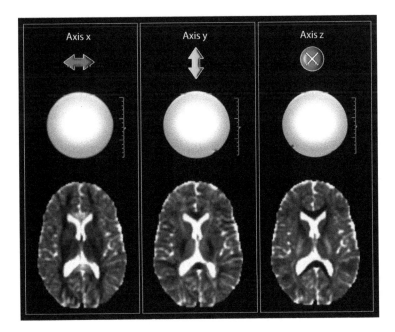

FIGURE 1.7   Three different DW images of a water phantom are depicted, one for each of the principal axes of the scanner, x, y, and z. The lower part of the figure depicts the DW images of the same three principal axis gradients of the scanner for a healthy volunteer.

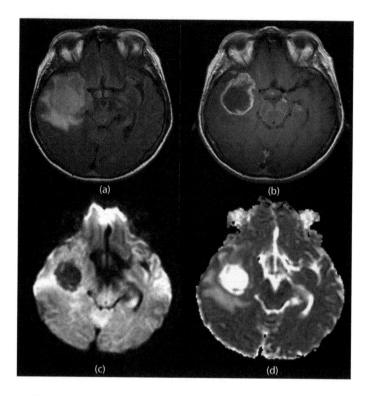

FIGURE 1.8   Axial T2-FLAIR (a) and T1-weighted post-contrast (b) images demonstrate a right temporal lesion with surrounding edema and ring-shaped enhancement. On the DW image, the lesion presents low signal intensity (c) resulting in higher intratumoral ADC (d).

1.2.5 Echo Planar Imaging

Since even minimal bulk patient motion during acquisition of DW images can obscure the effects of the much smaller microscopic water motion due to diffusion, fast imaging sequences are necessary for successful clinical DWI. The most widely used DW acquisition technique is single-shot echo-planar imaging (EPI). This is because in a clinical environment, certain requirements are imposed for diffusion studies. First, reasonable imaging time should be achieved (i.e., fast imaging). Second, multiple slices (15–20) are required to cover most of the brain, with good spatial resolution (~3–5 mm thick, 1–3 mm in-plane is required, at a reasonably short TE (120 ms) to reduce T2 decay, and an adequate diffusion sensitivity (ADC ~0.2–1 × 10^{-3} mm^2/s for brain tissues). The EPI sequence is fast and insensitive to small motion, which is essential. It is also readily available on most clinical MRI scanners. Because images can be acquired in a fraction of a second, artifacts from patient motion are greatly reduced, and motion between acquisitions with the different required diffusion-sensitizing gradients is also decreased.

Nevertheless, EPI suffers from limitations, which include the limited spatial resolution due to smaller imaging matrices as well as the blurring effect of T2* decay occurring during image readout. Other limitations are sensitivity to artifacts due to magnetic field inhomogeneity, chemical shift effects, ghosting, and local susceptibility effects. The latter is particularly important, as it results in marked distortion and signal drop-out near air cavities, particularly at the skull base and the posterior fossa, limiting sensitivity of DWI with EPI in these areas. Nonetheless, artifacts and pitfalls of DWI are going to be discussed in detail in the next chapter.

Alternative DWI techniques include multi-shot EPI with navigator echo correction or DW, periodically rotated overlapping parallel lines with enhanced reconstruction (PROPELLER) and parallel imaging methods, such as sensitivity encoding (SENSE) (Jones et al., 1999; Porter and Mueller, 2004). The application of such techniques increase the bandwidth per voxel in the phase encode direction, thus reducing artifacts arising from field inhomogeneities, like those induced by eddy currents and local susceptibility gradients.

1.2.6 Main Limitations of DWI

DWI is undoubtedly a very useful clinical tool and can help the visual interpretation of clinical images. However, it is only a qualitative type of exam, and is very sensitive to the choice of acquisition parameters and patient motion in the scanner.

Moreover, DWI sequences are sensitive, but not specific for the detection of restricted diffusion. Thus, one should not use only signal changes to quantify diffusion properties, as the signal from DWI is prone to the underlying T2-weighted signal, referred to as the "T2 shine-through" effect (Chilla et al. 2015). That is, the increased signal in areas of cytotoxic edema on T2-weighted images may be present on the DWI images as well (Jones et al., 1999). To determine whether this signal hyperintensity on DWI images truly represents decreased diffusion, the ADC map should also be used. The ADC sequence is not as sensitive as the DWI sequence for restricted diffusion, but it is more specific; the ADC images are not susceptible to the "T2 shine-through" effect since they are "relative" images (Debnam and Schellingerhout, 2011).

As described above, a typical clinical diffusion imaging protocol consists of four images at each level: (1) one without diffusion weighting (S_0), also known as the b = 0 s/mm^2 or "b zero" image, which has an image contrast similar to that of a conventional T2-weighted spin-echo image (for echo times and repetition times used in typical diffusion applications) and (2) three images with diffusion weighting along mutually orthogonal directions. For the reasons described earlier, the DW images the radiologist evaluates are not the set of orthogonally weighted images, but rather the geometric mean computed from these three images, also known as the isotropic DW image, or simply the ADC. Hence ADC is equal to

$$ADC = \frac{ADC1 + ADC2 + ADC3}{3} \tag{1.5}$$

where $ADC1$, $ADC2$, and $ADC3$ are the apparent diffusion coefficients along the directions of the three diffusion-sensitizing gradients. In terms of the acquired signal SDWI in the DW image, we have

$$S_{DWI} = (S_1 \times S_2 \times S_3)^{\frac{1}{3}} = S_0 e^{\left(-b \times \frac{ADC1 + ADC2 + ADC3}{3}\right)} = S_0 e^{(-b \times ADC)} \tag{1.6}$$

An important observation in this last term, is that there are two major sources of contrast in the DW image: the T2-weighted term S_0 and the exponential term related to diffusion. Hence, hyperintensity on DWI may be related to T2 prolongation (large S_0 term), reduced diffusion, or both. When high signal intensity is observed on DWI due to a dominant T2-related term in the setting of normal or even elevated ADC, it is known as T2 shine-through. Simply examining the b zero image or corresponding conventional T2-weighted image is not a reliable method for differentiating between truly reduced diffusion and T2 shine-through since both prolonged T2 and reduced diffusion may coexist (Pauleit et al., 2004). The T2 shine-through effect of a low-grade glioma is depicted in Figure 1.9.

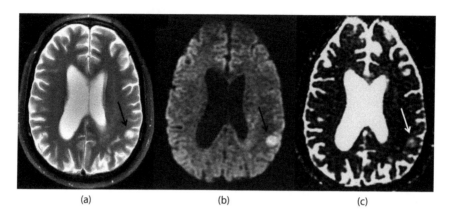

(a)   (b)   (c)

FIGURE 1.9 T2 "shine-through" effect of a low-grade glioma shown on a T2-weighted image (a), which appears bright on the DW image (b) and also bright on the ADC map (c), implying increased diffusivity. (Courtesy Allen D. Elster, MRIquestions.com.)

Unfortunately, ADC suffers from a limitation too. It depends on the direction of the applied diffusion encoding gradient, as was illustrated in Figure 1.7, where it is evident that in certain regions of the brain, ADC is different depending on the applied gradient. This effect, of course, enables us to extract valuable information about the brain microstructure; nevertheless, it also reveals that ADC is directionally dependent (Chenevert et al., 1990; Doran et al., 1990).

In other words, a single ADC would be inadequate for characterizing diffusion in clinical practice as it would depend on the direction of gradients as well as the position and possible movement of the patient's head. Although in clinical practice the average of the ADC values along the three orthogonal directions is used, known as the mean diffusivity, "trace," or, simply, the ADC, it is clear that an infinite number of ADC measures can be obtained within anisotropic tissue. This limitation was remedied by a more complex description as the diffusion tensor matrix, which is going to be discussed in detail in the next section.

1.3 Diffusion Tensor Imaging

Focus Point

- ADC is directionally dependent.
- A single ADC is inadequate for characterizing diffusion in vivo.
- Diffusion tensor imaging represents a further development of DWI.
- The diffusion tensor describes an ellipsoid that represents the directional movement of water molecules inside a voxel.

Diffusion tensor imaging (DTI) evolved from DWI and was developed to remedy the limitations of DWI (see previous section), taking advantage of the preferential water diffusion inside the brain tissue (Le Bihan, 2003; Mukherjee, et al., 2008). The water diffusion in the brain is NOT an isotropic process, due to the natural intracellular (neurofilaments and organelles) and extracellular (glial cells and myelin sheaths) barriers that restrict diffusion towards certain directions. Hence, water molecules diffuse mainly along the direction of white matter axons rather than perpendicular to them (please refer to Figure 1.1). Under these circumstances,

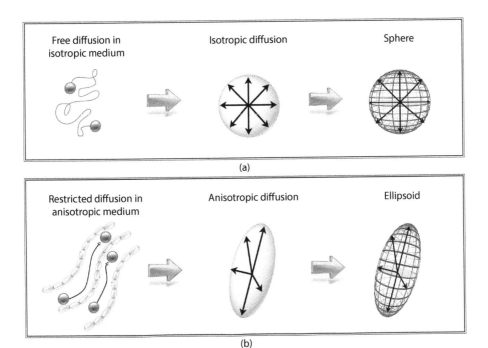

diffusion can become highly directional along the length of the tract, and is called anisotropic (Price, 2007) (Figure 1.10). This means that we talk about media that have different diffusion properties in different directions. In other words, in certain regions of the brain, ADC is directionally dependent; it is therefore also clear that a single ADC would be inadequate for characterizing diffusion and a more compound mathematical description is required.

In that that sense, DTI measures both the magnitude and the direction of proton movement within a voxel for multiple dimensions of movement using a mathematical model to represent this information, called the diffusion tensor (DT) (Debnam and Schellingerhout, 2011). Assuming that the probability of molecular displacements follows a multivariate Gaussian distribution over the observation diffusion time, the diffusion process can be described by a 3×3 tensor matrix, proportional to the variance of the Gaussian distribution. Thus, the diffusion tensor, D, is characterized by nine elements:

$$D = \begin{bmatrix} D_{xx} & D_{xy} & D_{xz} \\ D_{yx} & D_{yy} & D_{yz} \\ D_{zx} & D_{zy} & D_{zz} \end{bmatrix} \tag{1.7}$$

Now, consider that the directional movement of water molecules inside a voxel can be represented by an ellipsoid (Figure 1.11), which in turn can be described by the tensor in that specific voxel.

This tensor consists of the 3×3 matrix derived from diffusivity measurements in at least six different directions. This is because the tensor is diagonally symmetric ($D_{xy} = D_{yx}$, $D_{yz} = D_{zy}$, and $D_{xz} = D_{zx}$), therefore only six unknown elements need to be determined. Figure 1.12 shows the elements of the diffusion tensor. The images of D_{xx}, D_{yy}, and D_{zz}

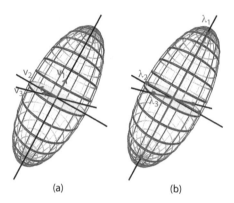

FIGURE 1.11 Three eigenvectors describe the orientation of the three axes of the ellipsoid (a), and three eigenvalues represent the magnitude of the axes (apparent diffusivities) of the ellipsoid (b).

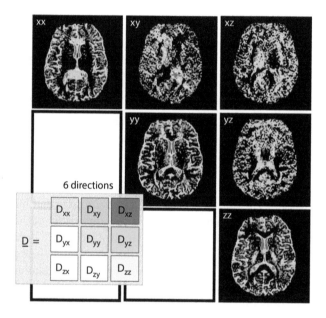

FIGURE 1.12 Elements of the diffusion tensor. The images of D_{xx}, D_{yy}, and D_{zz} show the diffusivity along the x-, y-, and z axes, respectively, while the images of D_{xy}, D_{xz}, and D_{yz} show respective displacements in orthogonal directions. Since the tensor is diagonally symmetric ($D_{xy} = D_{yx}$, $D_{yz} = D_{zy}$, and $D_{xz} = D_{zx}$), only six unknown elements need to be determined.

show the diffusivity along the x-, y-, and z axes, respectively, while the images of D_{xy}, D_{xz}, and D_{yz} show respective displacements in orthogonal directions.

If the tensor is completely aligned with the anisotropic medium, then the off-diagonal elements become zero and the tensor is diagonalized. This diagonalization provides three eigenvectors that describe the orientation of the three axes of the ellipsoid, and three eigenvalues, which represent the magnitude of the axes (apparent diffusivities) in the corresponding directions (Figure 1.12). The major axis is considered to be oriented in the direction of maximum diffusivity, which has been shown to coincide with tract orientation (Field and Alexander, 2004; Price, 2007). Therefore, there is a transition through the diffusion tensor from the x, y, z coordinate system defined by the scanner's geometry, to a new independent coordinate system, in which axes are dictated by the directional diffusivity information.

Depending on the local diffusion, the ellipsoid may be "prolate," "oblate," or "spherical." Prolate shapes are expected in highly organized tracts where the fiber bundles all have similar orientations, oblate shapes are expected when fiber orientations are more variable but remain limited to a single plane, and spherical shapes are expected in areas that allow isotropic diffusion (Alexander et al., 2000).

Going back to our example of ink in the glass of water, over time, the ink particles displace and, because the medium is isotropic, the outer surface of the displacements would resemble a sphere. On the contrary, if water was an anisotropic medium, the ink particles would diffuse preferentially along the principal axis of the anisotropic medium rather than perpendicular to it.

Then, the corresponding displacement profile can no longer be described by a sphere and is more correctly described by an ellipsoid, with the long axis parallel to the long axis of the anisotropic medium as depicted in Figure 1.10.

1.3.1 "Rotationally Invariant" Parameters (Mean Diffusivity and Fractional Anisotropy)

Using the tensor data, the local diffusion anisotropy can be quantified by the calculation of "rotationally invariant" parameters. The most commonly reported indices that can be calculated are the mean diffusivity (MD) or "Trace" and fractional anisotropy (FA). MD is the mean of the eigenvalues, and represents a directionally measured average of water diffusivity, whereas FA derives from the standard deviation of the three eigenvalues.

More analytically, the trace is the sum of the three diagonal elements of the diffusion tensor (i.e., $D_{xx} + D_{yy} + D_{zz}$), which can be shown to be equal to the sum of its three eigenvalues. The mean Trace (Trace/3) can be thought of as being equal to the averaged mean diffusivity.

The image of the MD (i.e., trace/3) is depicted in Figure 1.13, which is produced by the average of the ADC indices along the three orthogonal axes (as in Figure 1.7). This averaging produces an evident loss of contrast in parenchyma in the MD map. Nevertheless, Pierpaoli et al. (1996), showed that in the b-value range typically used in clinical studies (b<1500 s/mm²), the MD is fairly uniform throughout parenchyma at a value of about 0.7×10^{-3} mm²/s. This is advantageous in the sense that the effects of anisotropy do not confound the detection of diffusion abnormalities, such as acute ischemic lesions (Lythgoe et al., 1997; Lee et al., 2008). It should be evident, however, that if the b-value is changed, there will be dissociation between white and gray matter (Yoshiura et al., 2001), and moreover, it is obvious that in order to compare results between different institutions the b-value should be the same.

In the same logic, in order to specify a simple but unbiased anisotropy index, Pierpaoli and Basser (1996) came up with the FA and relative anisotropy (RA) indices. These are given by the following equations:

$$FA = \sqrt{\frac{3}{2}} \frac{\sqrt{\left(\lambda_1 - \langle\lambda\rangle\right)^2 + \left(\lambda_2 - \langle\lambda\rangle\right)^2 + \left(\lambda_3 - \langle\lambda\rangle\right)^2}}{\sqrt{\lambda_1^2 + \lambda_2^2 + \lambda_3^2}} \qquad (1.8)$$

and

$$RA = \sqrt{\frac{1}{2}} \frac{\sqrt{\left(\lambda_1 - \langle\lambda\rangle\right)^2 + \left(\lambda_2 - \langle\lambda\rangle\right)^2 + \left(\lambda_3 - \langle\lambda\rangle\right)^2}}{\langle\lambda\rangle} \qquad (1.9)$$

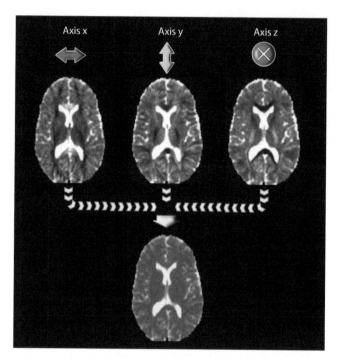

FIGURE 1.13   Average of the three ADC maps is the mean diffusivity (mathematically equivalent to one third of the trace of the diffusion tensor).

Both parameters indicate how elongated the diffusion ellipsoid is; hence, the information provided is essentially the same, although FA is the parameter most widely used. In an FA map, the signal brightness of a voxel, describes the degree of anisotropy in the given voxel. FA ranges from 0 to 1, depending on the underlying tissue architecture. A value closer to 0 indicates that the diffusion in the voxel is isotropic (unrestricted water movement), such as in areas of CSF, whereas a value closer to 1 describes a highly anisotropic medium, such as in the corpus callosum where water molecules diffuse along a single axis (Price, 2007). Example images showing FA for the whole brain in the axial plane are presented in Figure 1.14.

Diffusion directionality in various regions of interest can be further represented by a directionally encoded color (DEC) FA map as shown in Figure 15d. More specifically, the eigenvector with the largest eigenvalue defines the orientation of the ellipsoid in each voxel, which can then be color-coded to evaluate and display information about the direction of white matter tracts. Hence, ellipsoids describing diffusion from left to right are red (x-axis), ellipsoids describing anterioposterior (y-axis) diffusion are green, and diffusion in the cranio-caudal direction is blue (z-axis) (Pajevic and Pierpaoli, 1999). This procedure provides a user friendly and convenient summary map from which one can determine the degree of anisot-ropy (in terms of signal brightness) and the fiber orientation in the voxel (in terms of hue). A neuroradiologist can then combine and correlate this information with normal brain anatomy, identify specific white matter tracts, and assess the impact of a lesion on neigh-boring white matter fibers (Ferda et al., 2010). Figure 1.15 depicts the comparison of a T2-weighted, average diffusion coefficient (DC), fractional anisotropy (FA) map, and color-coded orientation map.

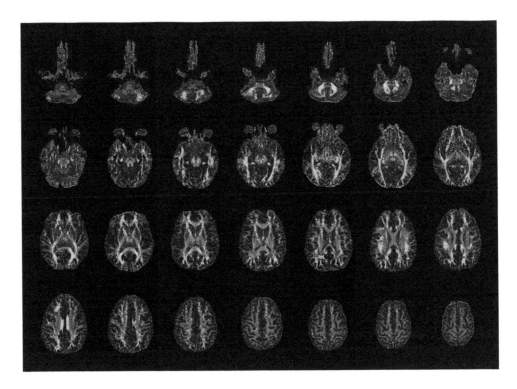

FIGURE 1.14 Example images showing FA for the whole brain in the axial plane. Directionally DEC FA map.

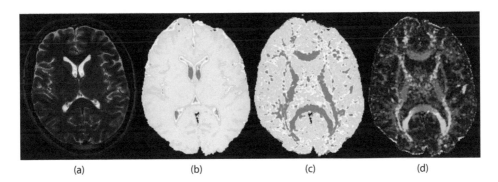

(a)   (b)   (c)   (d)

FIGURE 1.15 Comparison of T2-weighted (a), average diffusion coefficient (DC) (b), fractional anisotropy (FA) map (c), and color-coded orientation map (d). Images were acquired using a 3.0 T scanner. The colors represent the orientations of fibers; red: right–left, green: anterior–posterior, and blue: superior–inferior.

1.3.2 Fiber Tractography

By now, it must be evident that the underlying tissue structure dictates diffusion anisotropy, and in the human brain, this is mainly the white matter architecture. Hence, by combining FA values with directionality, it would be possible to obtain estimates of fiber orientation. This idea has led to the development of fiber tractography enabling the mapping of white matter tracts noninvasively (Westin et al., 2002; Assaf and Pasternak, 2008), as seen in Figure 1.16.

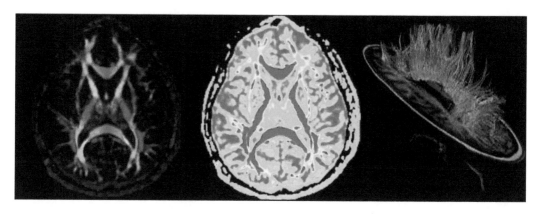

FIGURE 1.16   MR fiber tractography.

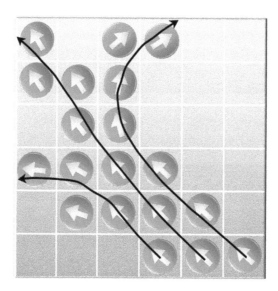

FIGURE 1.17   Schematic diagram of the line propagation approach.

Different algorithms have been developed for fiber tractography, but the main idea is that following the tensor's orientation on a voxel-by-voxel basis, it is possible to identify intravoxel connections and display specific fiber tracts using computer graphic techniques (Figure 1.17). A variety of tractography techniques have been reported (Jones et al., 1999; Mori et al., 1999; Mori et al., 2002, Parker et al., 2002). All these techniques use mathematical models to identify neighboring voxels that might be located within the same fiber tract based on the regional tensor orientations and relative positions of the voxels.

Towards this direction, a number of studies have created atlases of the human brain based on DTI and tractography (Jellison et al., 2004; Wakana et al., 2004, Mori et al., 2009). According to this, a very important differential diagnostic parameter regarding the displacement or disruption of a specific fiber tract by a pathology may be assessed by 3D tractograms (Bello et al., 2010; Mori and Zijl, 2002), as is displayed in a case of brain tumor tractography in Figure 1.18.

In order to produce tracts, the user needs to define a "seed" region of interest (ROI) on the color orientation map that is very useful in visualizing the white matter tract orientation. In most software applications, this is defined as "Structural View." Depositing a seed ROI results in a white matter track oriented through the ROI. To display white matter tracks oriented from one ROI to another ROI, one needs to position a second ROI on the image and define a "target" ROI. This is illustrated in Figure 1.19.

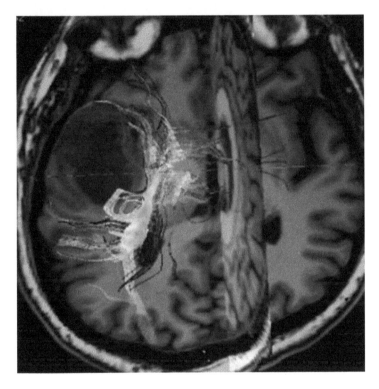

FIGURE 1.18 Displacement or disruption of a specific fiber tract by a pathology may be assessed by 3D tractograms.

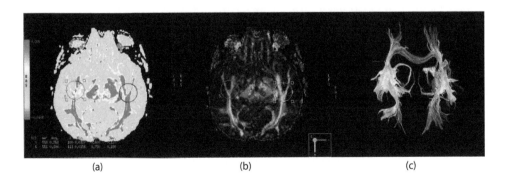

(a)   (b)   (c)

FIGURE 1.19 FA map (a), ROI placement on colored orientation map (b), and fiber tracts (c).

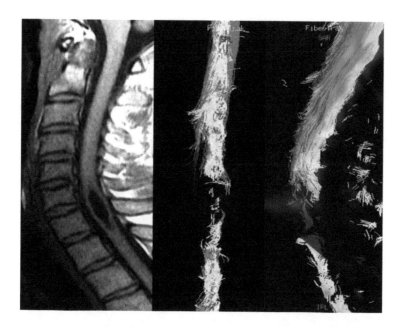

FIGURE 1.20   Spinal cord diffusion tensor tractography. (Courtesy of General Electric, with permission, and Mt. Sinai Hospital.)

These techniques also provide useful information in terms of presurgical planning (Romano et al., 2009; Arfanakis et al., 2006). Nonetheless, they present limitations such as in cases of complex tracts (crossing or branching fibers), which should be taken into consideration when these methods are used for preoperative guidance (Jones, 2010).

In DTI, the diffusion gradients are applied in multiple directions, and based on previous reports, the number of non-collinear gradients applied varies (ranging from 6 to 55). There is much debate in the literature; however, an optimal number has not yet been defined (Hasan et al., 2001; Jones, 2004; Nucifora et al., 2007). As one can imagine, the main drawback of an increased number of gradients in DTI is the imaging time, which increases simultaneously and may not be applicable in clinical practice (Gupta et al., 2010). Therefore, as always, there is a trade-off between the imaging time and the number of gradients applied in order to obtain sufficient diffusion information.

DTI has also been applied in the spinal cord, in the evaluation of acute and chronic trauma, tumors of the spinal canal, degenerative myelopathy, demyelinating and infectious diseases, and so on, and there are strong indications that it can be a sensitive and specific method (Jones et al., 1999). Figure 1.20 illustrates a case of spinal cord diffusion tensor tractography.

It has to be noted that there are still many technical limitations in the application of spinal cord DTI, especially in thoracic and lumbar segments. Nevertheless, the wider use of higher field scanners (3T or more), and the further development of acquisition and post-processing techniques, should result in the increased role of this promising advanced technique in both research and clinical practice.

1.4 Conclusions and Future Perspectives

The usefulness of conventional MRI in the detection of cerebral pathology has been well-established, although it can be in many cases nonspecific despite the excellent soft tissue visualization. The addition of DWI and DTI has truly revolutionized clinical neuroimaging, providing microstructural information with specific benefits, which can be summarized in the following points:

1. Pathology may be detected earlier and in a quantitative manner, allowing increased specificity.
2. The microarchitecture of the brain can now be deeply explored.
3. DWI/DTI metrics can be used as quantifiable objective features allowing tumor classification as well as treatment monitoring.
4. Diffusion may aid the differentiation between cytotoxic brain edema (restricted diffusion) and vasogenic edema (increased diffusion) offering both diagnostic (tumor categorization) and prognostic (reversible pathology) value.
5. The functional connectivity within the brain is now explored using DTI to evaluate white matter tracts. Besides clinical studies, this is expected to optimize surgery planning and therefore treatment outcome.

Moreover, a number of significant future clinical applications will emerge, as there is intensive ongoing research in the field with increasing applications, which will be translated into routine clinical neuroimaging. Based on the aforementioned points, it is expected that the diagnosis of several pathologies such as ischemia, infection, and demyelinating disease will benefit as well.

Nevertheless, caution against over-reliance on "scientific extras" and "advanced tools" is needed since we must never forget that even as sophisticated a mathematic construct as the DTI model is, it is an oversimplification of the properties of water diffusion in the brain, with several associated limitations.

These limitations mainly involve the complex white matter architecture with kissing, branching and intersecting fiber tracts, which may result in erroneous estimation of the white matter tracks, as well as in problematic evaluation of diffusion indices like FA or MD.

The limitations of DWI/DTI techniques with their associated artifacts and pitfalls that one should take into account are going to be analytically discussed in the next chapter.

References

Alexander, A. L., Hasan, K., Kindlmann, G., Parker, D. L., and Tsuruda, J. S. (2000). A geometric analysis of diffusion tensor measurements of the human brain. *Magnetic Resonance in Medicine*, 44(2), 283–291. doi:10.1002/1522-2594(200008)44:2<283::aid-mrm16>3.0.co;2-v

Arfanakis, K., Gui, M. and Lazar, M. (2006). Optimization of white matter tractography for pre-surgical planning and image-guided surgery. *Oncology Reports*, 15 Spec no., 1061–1064. 10.3892/or.15.4.1061.

Assaf, Y. and Basser, P. J. (2005). Composite hindered and restricted model of diffusion (CHARMED) MR imaging of the human brain. *NeuroImage*, 27(1), 48–58. doi:10.1016/j.neuroimage.2005.03.042">10.1016/j.neuroimage.2005.03.042

Assaf, Y. and Pasternak, O. (2008). Diffusion tensor imaging (DTI)-based white matter mapping in brain research: A review. *Journal of Molecular Neuroscience*, 34(1), 51–61. https://doi.org/10.1007/s12031-007-0029-0

Beaulieu, C. (2002). The basis of anisotropic water diffusion in the nervous system—A technical review. *NMR in Biomedicine*, 15(7–8), 435–455. doi:10.1002/nbm.782

Beaulieu, C. and Allen, P. S. (1994a). Determinants of anisotropic water diffusion in nerves. *Magnetic Resonance in Medicine*, 31(4), 394–400. doi:10.1002/mrm.1910310408

Beaulieu, C. and Allen, P. S. (1994b). Water diffusion in the giant axon of the squid: Implications for diffusion-weighted MRI of the nervous system. *Magnetic Resonance in Medicine*, 32(5), 579–583. doi:10.1002/mrm.1910320506

Beaulieu, C. and Allen, P. S. (1996). An in vitro evaluation of the effects of local magnetic-susceptibility-induced gradients on anisotropic water diffusion in nerve. *Magnetic Resonance in Medicine*, 36(1), 39–44. doi:10.1002/mrm.1910360108

Bello, L., Castellano, A., Fava, E., Casaceli, G., Riva, M., Scotti, G. et al. (2010). Intraoperative use of diffusion tensor imaging fiber tractography and subcortical mapping for resection of gliomas: Technical considerations. *Neurosurgical Focus*, 28(2). doi:10.3171/2009.12.focus09240

Ben-Amitay, S., Jones, D. K., and Assaf, Y. (2012). Motion correction and registration of high b-value diffusion weighted images. *Magnetic Resonance in Medicine*, 67(6), 1694–1702. doi:10.1002/mrm.23186

Bloch, F. (1950). Nuclear induction. *Physics Today*, 3(8), 22–25. doi:10.1063/1.3066970

Chenevert, T. L., Brunberg, J. A., and Pipe, J. G. (1990). Anisotropic diffusion in human white matter: Demonstration with MR techniques in vivo. *Radiology*, 177(2), 401–405. doi:10.1148/radiology.177.2.2217776

Chilla, G. S., Tan, C. H., Xu, C., and Poh, C. L. (2015). Diffusion weighted magnetic resonance imaging and its recent trend—A survey. *Quantitative Imaging in Medicine and Surgery*, 5(3), 407–422. http://doi.org/10.3978/j.issn.2223-4292.2015.03.01

Debnam, J. M. and Schellingerhout, D. (2011). Diffusion MR imaging of the brain in patients with cancer. *International Journal of Molecular Imaging*, 2011, 1–9. doi:10.1155/2011/714021

Doran, M., Hajnal, J. V., Bruggen, N. V., King, M. D., Young, I. R., and Bydder, G. M. (1990). Normal and Abnormal White Matter Tracts Shown by MR Imaging using Directional Diffusion Weighted Sequences. *Journal of Computer Assisted Tomography*, 14(6), 865–873. doi:10.1097/00004728-199011000-00001

Einstein, A. (1905). Über die von der molekularkinetischen Theorie der Wärme geforderte Bewegung von in ruhenden Flüssigkeiten suspendierten Teilchen. *Annalen der Physik*, 322(8), 549–560. doi:10.1002/andp.19053220806

Fan, G. G., Deng, Q. L., Wu, Z. H., and Guo, Q. Y. (2006). Usefulness of diffusion/perfusion-weighted MRI in patients with non-enhancing supratentorial brain gliomas: A valuable tool to predict tumour grading? *The British Journal of Radiology*, 79(944), 652–658. doi:10.1259/bjr/25349497

Ferda, J., Kastner, J., Mukenšnabl, P., Choc, M., Horemužová, J., Ferdová, E., and Kreuzberg, B. (2010). Diffusion tensor magnetic resonance imaging of glial brain tumors. *European Journal of Radiology*, 74(3), 428–436. doi:10.1016/j.ejrad.2009.03.030

Field, A. S. and Alexander, A. L. (2004). Diffusion tensor imaging in cerebral tumor diagnosis and therapy. *Topics in Magnetic Resonance Imaging*, 15(5), 315–324. doi:10.1097/00002142-200410000-00004

Gillard, J. H., Waldman, A. D., and Barker, P. B. (2005). *Clinical MR Neuroimaging: Physiological and functional techniques*. Cambridge: Cambridge University Press.

Guo, A. C., Macfall, J. R., and Provenzale, J. M. (2002). Multiple sclerosis: Diffusion tensor MR imaging for evaluation of normal-appearing white matter. *Radiology*, 222(3), 729–736. doi:10.1148/radiol.2223010311

Gupta, A., Holodny, A. I., Shah, A., and Young, R. J. (2010). Imaging of brain tumors: Functional magnetic resonance imaging and diffusion tensor imaging. *Neuroimaging Clinics of North America*, 20(3), 379–400.

Hahn, E. L. (1950). Spin echoes. *Physical Review*, 80(4), 580–594. doi:10.1103/physrev.80.580

Hakyemez, B., Erdogan, C., Gokalp, G., Dusak, A., and Parlak, M. (2010). Solitary metastases and high-grade gliomas: Radiological differentiation by morphometric analysis and perfusion-weighted MRI. *Clinical Radiology*, 65(1), 15–20. doi:10.1016/j.crad.2009.09.005

Hasan, K. M., Parker, D. L., and Alexander, A. L. (2001). Comparison of gradient encoding schemes for diffusion-tensor MRI. *Journal of Magnetic Resonance Imaging*, 13(5), 769–780. doi:10.1002/jmri.1107

Hong, X. and Dixon, W. T. (1992). Measuring diffusion in inhomogeneous systems in imaging mode using antisymmetric sensitizing gradients. *Journal of Magnetic Resonance (1969)*, 99(3), 561–570. doi:10.1016/0022-2364(92)90210-x

Jellison, B. J., Alexander, A. L., Field, A. S., Lazar, M., Medow, J., and Salamat, M. S. (2004). Diffusion tensor imaging of cerebral white matter: A pictorial review of physics, fiber tract anatomy, and tumor imaging patterns. *AJNR American journal of neuroradiology, 25*(3), 356–369.

Jones, D., Horsfield, M., and Simmons, A. (1999). Optimal strategies for measuring diffusion in anisotropic systems by magnetic resonance imaging. *Magnetic Resonance in Medicine, 42*(3), 515–525. doi:10.1002/(SICI)1522-2594(199909)42:3<515::AID-MRM14>3.0.CO;2-Q

Jones, D. K. (2004). The effect of gradient sampling schemes on measures derived from diffusion tensor MRI: A Monte Carlo study. *Magnetic Resonance in Medicine, 51*(4), 807–815. doi:10.1002/mrm.20033

Jones, D. K. (2010). Challenges and limitations of quantifying brain connectivity in vivo with diffusion MRI. *Imaging in Medicine, 2*(3), 341–355. doi:10.2217/iim.10.21

Jones, D. K., Knösche, T. R., and Turner, R. (2013). White matter integrity, fiber count, and other fallacies: The do's and don'ts of diffusion MRI. *NeuroImage, 73*, 239–254. doi:10.1016/j.neuroimage.2012.06.081

Jones, D. K., Simmons, A., Williams, S. C., and Horsfield, M. A. (1999). Non-invasive assessment of axonal fiber connectivity in the human brain via diffusion tensor MRI. *Magnetic Resonance in Medicine, 42*(1), 37–41. doi:10.1002/(sici)1522-2594(199907)42:1<37::aid-mrm7>3.0.co;2-o

Kaden, E., Kelm, N. D., Carson, R. P., Does, M. D., & Alexander, D. C. (2016). Multi-compartment microscopic diffusion imaging. *NeuroImage, 139*, 346–359. doi:10.1016/j.neuroimage.2016.06.002

Kono, K., Inoue, Y., Morino, M., Nakayama, K., Ohata, K., Shakudo, M., Wakasa, K., and Yamada, R. (2001). The role of diffusion-weighted imaging in patients with brain tumors. *AJNR American Journal of Neuroradiology, 22*(6), 1081–1088.

Lam, W., Poon, W., and Metreweli, C. (2002). Diffusion MR imaging in glioma: Does it have any role in the pre-operation determination of grading of glioma? *Clinical Radiology, 57*(3), 219–225. doi:10.1053/crad.2001.0741

Le Bihan, D. (2003). Looking into the functional architecture of the brain with diffusion MRI. *Nature Reviews Neuroscience, 4*(6), 469–480. doi:10.1038/nrn1119

Le Bihan, D., Breton, E., Lallemand, D., Grenier, P., Cabanis, E., and Laval-Jeantet, M. (1986). MR imaging of intravoxel incoherent motions: Application to diffusion and perfusion in neurologic disorders. *Radiology, 161*(2), 401–407. doi:10.1148/radiology.161.2.3763909

Le Bihan, D., Poupon, C., Amadon, A., and Lethimonnier, F. (2006). Artifacts and pitfalls in diffusion MRI. *Journal of Magnetic Resonance Imaging, 24*(3), 478–488. doi:10.1002/jmri.20683

Lee, H. Y., Na, D. G., Song, I., Lee, D. H., Seo, H. S., Kim, J., and Chang, K. (2008). Diffusion-tensor imaging for glioma grading at 3-T magnetic resonance imaging. *Journal of Computer Assisted Tomography, 32*(2), 298–303. doi:10.1097/rct.0b013e318076b44d

Lian, J., Williams, D., and Lowe, I. (1994). Magnetic resonance imaging of diffusion in the presence of background gradients and imaging of background gradients. *Journal of Magnetic Resonance, Series A, 106*(1), 65–74. doi:10.1006/jmra.1994.1007

Lythgoe, M. F., Busza, A. L., Calamante, F., Sotak, C. H., King, M. D., Bingham, A. C., and Gadian, D. G. (1997). Effects of diffusion anisotropy on lesion delineation in a rat model of cerebral ischemia. *Magnetic Resonance in Medicine, 38*(4), 662–668. doi:10.1002/mrm.1910380421

Mori, S., Crain, B. J., Chacko, V. P., and Zijl, P. C. (1999). Three-dimensional tracking of axonal projections in the brain by magnetic resonance imaging. *Annals of Neurology, 45*(2), 265–269. doi:10.1002/1531-8249(199902)45:2<265::aid-ana21>3.0.co;2-3

Mori, S., Frederiksen, K., Zijl, P. C., Stieltjes, B., Kraut, M. A., Solaiyappan, M., and Pomper, M. G. (2002). Brain white matter anatomy of tumor patients evaluated with diffusion tensor imaging. *Annals of Neurology, 51*(3), 377–380. doi:10.1002/ana.10137

Mori, S., Oishi, K., and Faria, A. V. (2009). White matter atlases based on diffusion tensor imaging. *Current Opinion in Neurology, 22*(4), 362–369. http://doi.org/10.1097/WCO.0b013e32832d954b

Mori, S. and Zijl, P. C. (2002). Fiber tracking: Principles and strategies—A technical review. *NMR in Biomedicine, 15*(7–8), 468–480. doi:10.1002/nbm.781

Moritani, T., Ekholm, S., and Westesson, P. (2009). *Diffusion-Weighted MR Imaging of the Brain*. Berlin: Springer.

Mukherjee, P., Berman, J., Chung, S., Hess, C., and Henry, R. (2008). Diffusion tensor MR imaging and fiber tractography: Theoretic underpinnings. *American Journal of Neuroradiology*, 29(4), 632–641. doi:10.3174/ajnr.a1051

Nagar, V., Ye, J., Ng, W., Chan, Y., Hui, F., Lee, C., and Lim, C. (2008). Diffusion-weighted MR imaging: diagnosing atypical or malignant meningiomas and detecting tumor dedifferentiation. *American Journal of Neuroradiology*, 29(6), 1147–1152. doi:10.3174/ajnr.a0996

Nucifora, P. G., Verma, R., Lee, S., and Melhem, E. R. (2007). Diffusion-tensor MR imaging and tractography: Exploring brain microstructure and connectivity. *Radiology*, 245(2), 367–384. doi:10.1148/radiol.2452060445

Pajevic, S. and Pierpaoli, C. (1999). Color schemes to represent the orientation of anisotropic tissues from diffusion tensor data: Application to white matter fiber tract mapping in the human brain. *Magnetic Resonance in Medicine*, 42(3), 526–540. doi:10.1002/(sici)1522-2594(199909)42:3<526::aid-mrm15>3.3.co;2-a

Parker, G. J., Stephan, K. E., Barker, G. J., Rowe, J. B., Macmanus, D. G., Wheeler-Kingshott, C. A. et al. (2002). Initial demonstration of in vivo tracing of axonal projections in the macaque brain and comparison with the human brain using diffusion tensor imaging and fast marching tractography. *NeuroImage*, 15(4), 797–809. doi:10.1006/nimg.2001.0994

Pauleit, D., Langen, K., Floeth, F., Hautzel, H., Riemenschneider, M. J., Reifenberger, G., and Müller, H. (2004). Can the apparent diffusion coefficient be used as a noninvasive parameter to distinguish tumor tissue from peritumoral tissue in cerebral gliomas? *Journal of Magnetic Resonance Imaging*, 20(5), 758–764. doi:10.1002/jmri.20177

Pierpaoli, C. and Basser, P. J. (1996). Toward a quantitative assessment of diffusion anisotropy. *Magnetic Resonance in Medicine*, 36(6), 893–906. doi:10.1002/mrm.1910360612

Pierpaoli, C., Jezzard, P., Basser, P. J., Barnett, A., and Chiro, G. D. (1996). Diffusion tensor MR imaging of the human brain. *Radiology*, 201(3), 637–648. doi:10.1148/radiology.201.3.8939209

Porter, D. and Mueller, E. (2004). Multi-shot diffusion-weighted EPI with readout mosaic segmentation and 2D navigator correction. *Proceedings of the 12th Annual Meeting of ISMRM*, 11, 442.

Price, S. J. (2007). The role of advanced MR imaging in understanding brain tumour pathology. *British Journal of Neurosurgery*, 21(6), 562–575. doi:10.1080/02688690701700935

Romano, A., D'Andrea, G., Minniti, G., Mastronardi, L., Ferrante, L., Fantozzi, L. M., and Bozzao, A. (2009). Pre-surgical planning and MR-tractography utility in brain tumour resection. *European Radiology*, 19(12), 2798–2808. doi:10.1007/s00330-009-1483-6

Schellinger, P. D., Fiebach, J. B., Jansen, O., Ringleb, P. A., Mohr, A., Steiner, T. et al. (2001). Stroke magnetic resonance imaging within 6 hours after onset of hyperacute cerebral ischemia. *Annals of Neurology*, 49(4), 460–469. doi:10.1002/ana.95.abs

Schmainda, K. M. (2012). Diffusion-weighted MRI as a biomarker for treatment response in glioma. *CNS Oncology*, 1(2), 169–180. doi:10.2217/cns.12.25

Seo, H., Chang, K., Na, D., Kwon, B., and Lee, D. (2008). High b-value diffusion (b = 3000 s/mm2) MR imaging in cerebral gliomas at 3T: Visual and quantitative comparisons with b = 1000 s/mm2. *American Journal of Neuroradiology*, 29(3), 458–463. doi:10.3174/ajnr.a0842

Stejskal, E. O. and Tanner, J. E. (1965). Spin diffusion measurements: Spin echoes in the presence of a time-dependent field gradient. *Journal of Chemical Physics*, 42(1), 288–292. doi:10.1063/1.1695690

Tanner, J. E. (1978). Transient diffusion in a system partitioned by permeable barriers. Application to NMR measurements with a pulsed field gradient. *Journal of Chemical Physics*, 69(4), 1748–1754. doi:10.1063/1.436751

Thomsen, C., Henriksen, O., and Ring, P. (1987). In vivo measurement of water self diffusion in the human brain by magnetic resonance imaging. *Acta Radiologica*, 28(3), 353–361. doi:10.1177/028418518702800324

Toh, C., Castillo, M., Wong, A., Wei, K., Wong, H., Ng, S., and Wan, Y. (2008b). Primary cerebral lymphoma and glioblastoma multiforme: Differences in diffusion characteristics evaluated with diffusion tensor imaging. *American Journal of Neuroradiology, 29*(3), 471–475. doi:10.3174/ajnr.a0872

Wakana, S., Jiang, H., Nagae-Poetscher, L. M., Zijl, P. C., and Mori, S. (2004). Fiber tract–based atlas of human white matter anatomy. *Radiology, 230*(1), 77–87. doi:10.1148/radiol.2301021640

Westin, C., Maier, S., Mamata, H., Nabavi, A., Jolesz, F., and Kikinis, R. (2002). Processing and visualization for diffusion tensor MRI. *Medical Image Analysis, 6*(2), 93–108. doi:10.1016/s1361-8415(02)00053-1

Yamasaki, F., Kurisu, K., Satoh, K., Arita, K., Sugiyama, K., Ohtaki, M. et al. (2005). Apparent diffusion coefficient of human brain tumors at MR imaging. *Radiology, 235*(3), 985–991. doi:10.1148/radiol.2353031338

Yoshiura, T., Wu, O., Zaheer, A., Reese, T. G., and Sorensen, A. G. (2001). Highly diffusion-sensitized MRI of brain: Dissociation of gray and white matter. *Magnetic Resonance in Medicine, 45*(5), 734–740. doi:10.1002/mrm.1100

Zakaria, R., Das, K., Radon, M., Bhojak, M., Rudland, P. R., Sluming, V., and Jenkinson, M. D. (2014). Diffusion-weighted MRI characteristics of the cerebral metastasis to brain boundary predicts patient outcomes. *BMC Medical Imaging, 14*(1), 26. doi:10.1186/1471-2342-14-26

2

Artifacts and Pitfalls in Diffusion MRI

2.1 Introduction

Focus Point

- Diffusion imaging experiments are sensitive to phenomena other than diffusion.
- DWI artifacts may be related to the scanner hardware, e.g., eddy currents, b-factor dependence, etc.
- Artifacts may be related to properties of the subject being imaged, e.g., T2 effects, motion artifacts, "physiological" noise.
- Artifacts and pitfalls are produced by processing and interpretation errors, e.g., local misalignment, heterogeneous voxels, etc.

Over the last 20 years, diffusion-weighted imaging (DWI) and more recently diffusion tensor imaging (DTI) techniques have become established tools in the clinical routine, with great impact on neurosciences regarding brain tissue integrity and especially white matter microstructure.

Nevertheless, the acquisition as well as the interpretation of diffusion imaging results are not always straightforward. This is both due to the sensitivity of diffusion imaging experiments to phenomena other than diffusion (pitfalls) (Norris, 2001), and to the fact that, like any other magnetic resonance imaging (MRI) technique, it remains subject to artifacts, numerous technical difficulties, and other sources of error.

More specifically, in addition to common MRI artifacts, DWI encounters characteristic problems regarding gradient hardware, especially in terms of eddy currents and sensitivity to motion; furthermore, there are complications regarding the analysis of the vast numbers of numerical data that may arise from a diffusion experiment, such as apparent diffusion coefficient, eigenvectors, eigenvalues, trace of the diffusion tensor, fractional anisotropy, etc.

It is true that considerable progress has been recently made, especially in overcoming hardware limitations and methodological shortcomings (Le Bihan et al., 2006). However, obtaining reliable data and interpreting accurate diffusion measurements remain challenging and is subject to the aforementioned problems.

However, the scope of this chapter is not to convey the idea that DWI might be unreliable, but on the contrary to introduce and analyze the artifacts and pitfalls in a qualitative way, as well as to evaluate possible mitigating strategies or solutions to overcome them. In that sense, the potential user of DWI techniques will be able to avoid or correct errors and make sound interpretations.

2.2 Artifacts and Pitfalls Categorization

The potential artifacts and pitfalls in DWI can emerge from a number of sources of error. These error sources may be divided into three distinct categories, according to whether they arise: (1) from the scanner hardware, (2) due to properties of the subject being imaged, or (3) from processing and interpretation pitfalls. The first two are not user dependent, nevertheless, a detailed review should help users recognize and mitigate them with post-processing techniques. The third category is more demanding, as potential errors may limit the accuracy and reliability of results and lead to substantial pitfalls. Many researchers of the field believe that the Holy Grail of DWI data processing is to reduce the margin of error as much as possible by identifying problems of the whole analysis pipeline, starting from acquisitions and hardware to final evaluation and interpretation (Jones and Cercignani, 2010; Jones et al., 2013; Le Bihan et al., 2006; Vavrek and MacFall, 1995).

Hence, the aforementioned sources of error can be divided into the following categories:

a. *Artifacts from the scanner's gradient system*
b. *Motion artifacts*
c. *Artifacts due to properties of the subject being imaged (i.e., physiological noise)*
d. *Processing and interpretation pitfalls*

2.3 Artifacts from the Gradient System

The most important parameter regarding DWI quality is the b-factor. The b-factor is produced by a combination of the strength and the duration of the gradient pulses that are necessary to detect and visualize the microscopic molecular displacements of water molecules (please refer to Chapter 1, Section 1.1.2). For the b-factor to be accurate and effective, there is a need for stable gradients of the utmost intensity, and this requirement may be extremely challenging when considering whole-body instruments designed for clinical studies (Vavrek and MacFall, 1995).

2.3.1 Eddy Current Artifacts

The main source of artifacts from the gradient system comes from the induced eddy currents. In terms of electromagnetism, eddy currents are loops of electrical current induced within conductors by a changing magnetic field in the conductor.

The high-performance gradient coils of modern MRI scanners produce rapidly switched magnetic field gradients during a pulse sequence. These in turn induce eddy currents in the electrically conductive structures of the scanner, which produce additional unwanted magnetic fields. These additional magnetic field gradients may interfere with the imaging gradients as they decay slowly and may persist after the primary gradients are switched off. This effect on the actual shape of the diffusion gradient pulses is illustrated in Figure 2.1.

More specifically, these eddy current-induced magnetic field gradients can either add to or subtract from the gradients that are used for spatial encoding. In "conventional" imaging, the gradients are normally applied only for short periods and possible induced eddy currents tend to self-cancel. The problem emerges in DWI, where there is a need for robust b-values of the order of 1000–3000 s/mm^2 using clinical scanners that generally have limited gradient power. To compensate for the limited power, gradients are applied for much longer. Thus, the rising and falling parts of the gradient waveform are sufficiently temporally separated and the eddy currents are no longer self-cancelled (Jones and Cercignani, 2010).

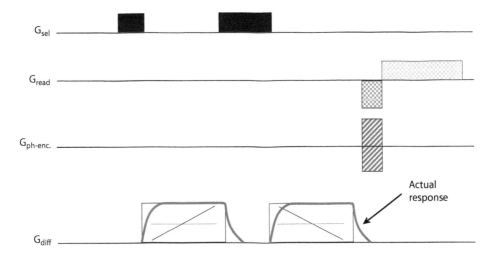

FIGURE 2.1 Effects of eddy currents on the actual shape of the diffusion gradient pulses. It is evident that eddy current-induced gradients interfere with the readout gradient.

In turn, the eddy currents generate magnetic field gradients that vectorially combine with the imaging gradient pulses, such that the actual gradients experienced by spins in the imaged objects are not exactly the same as those that were programmed to produce and reconstruct the image as seen in Figure 2.1 (Le Bihan et al., 2006). This error in the local gradients causes geometrical distortion of the final image. There are three specific patterns of image distortion that can be observed in the diffusion images affected by eddy currents: contraction, shift, and shear. Each one depends on the diffusion encoding applied. For example, if there is a residual eddy current along the slice-direction, every voxel in the slice will be affected by the same phase shift, so there will be a shift of the image slice along the phase-encoding direction. It is evident that those distortions are expected to become worse with the increase of the strength and duration of gradient pulses (i.e., the b-factor). In other words, eddy current distortions are b-factor dependent, and since the b-factor scales with the square of the gradient strength, the effect may be far from negligible (Le Bihan et al., 2006; Wright et al., 2007).

This is obviously very important since DWI is a quantitative technique in the sense that the diffusion coefficient or diffusion tensor is calculated in each voxel from a collection of DWIs, obviously assuming that the gradients actually being applied to the tissue are the same as the prescribed gradients.

Fortunately, recognizing such distortions is easy, e.g., by comparing or fusing the diffusion images with artifact-free anatomical images, or less distorted low-b-value images. The latter can also be used as reference to correct for eddy current distortions. A typical image of the artifact that is produced from uncorrected eddy current-induced distortions is a rim of high anisotropy along the phase-encoded direction, as seen in Figure 2.2.

Another important, though underestimated, source of inaccuracy is that the actual voxel size of the image might vary because of the mismatch between the actual local and prescribed gradients, so the calculated apparent diffusion coefficient (ADC) maps may be overestimated or underestimated.

The take-home message from this pitfall should be that when there is an obvious eddy current artifact on the image, it has to be taken into account. The notion that it is safe to collect measurements as long as one performs analyses away from the rim artefact is just incorrect.

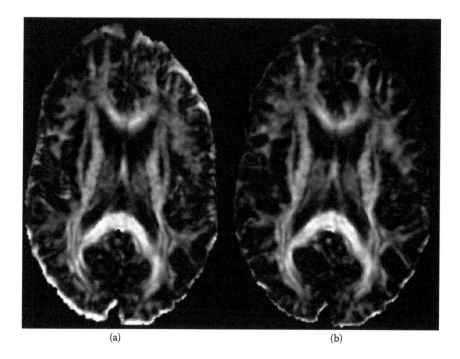

(a)   (b)

FIGURE 2.2   Two diffusion tensor images (relative diffusion anisotropy maps) that illustrate the eddy current artifact and the effectiveness of the correction technique; (a) without correction; (b) with correction.

2.3.2 Eddy Currents—Mitigating Strategies

As always, prevention is better than treatment, so first, artifacts must be recognized by using calibrated phantoms with a range of alkanes (cyclic and linear) to characterize both precision and accuracy (Tofts et al., 2000). The test should be repeated along the three physical axes of the scanner since eddy currents are different in the direction of gradient pulses. In an isotropic phantom, the ADC must be the same in all directions (see Chapter 1, Figure 1.7). Any differentiation should indicate either eddy currents or nonlinearity of the gradient system.

A number of approaches have been developed for minimizing eddy current artifacts. Early approaches aimed at correcting slices on an individual basis (since eddy current distortions are time varying and slices are acquired at different times) by matching the non-diffusion-weighted (non-DW) images, using mutual information-based cost functions or iterative cross-correlation (Bastin, 1999; Haselgrove and Moore, 1996). However, residual distortions would almost always persist, so vendors aimed at correcting eddy currents at the source by using so-called "self-shielded" gradient coils, which are now standard in every MRI system.

Nevertheless, residual currents may have remained, so an alternative approach included acquiring a reference scan in the absence of phase encoding and using this to determine the time-dependent phase shifts (Jezzard et al., 1998). Another approach was to use a mapping of the eddy current-induced fields in a separate experiment by directly imaging a phantom (Bastin and Armitage, 2000; Horsfield, 1999). Correction could then be achieved by acquiring one-dimensional field maps in the read and phase encoding direction for each slice and each diffusion step. By using the eddy current correction scheme outlined, the eddy current-induced artifacts in the DW images are almost completely eliminated. In addition, there is a significant improvement in the quality of the resulting diffusion tensor maps.

> Unfortunately, the previously described methods might in turn give rise to pitfalls.

First, the phantom correction method assumes there is no change in the gradient system between the two separate measurements, which might not be the case. Moreover, all the approaches mentioned previously assume that the eddy current distortions are uniform across all slices in a given volume, and that there is no subject motion, which again might not be the case. Last, one more obstacle to implementation is the difficulty of rapidly acquiring reliable field maps.

Hence, instead of the field map-based correction scheme, a more complete solution would be the use of a model that describes how eddy currents evolve, temporally and spatially, together with subject motion. Haselgrove and Moore (1996) first proposed the image-based registration method to correct for eddy current-induced distortions, while Rohde et al. (2004) proposed a novel and comprehensive approach. They aimed to correct for spatial misalignment of DWI volumes acquired with different strengths and orientations of the diffusion sensitizing gradients and to correct for eddy current-induced image distortion as well as rigid body motion in three dimensions as an optimization procedure.

Last, recent studies show that for optimizing the registration accuracy, it is more important to correct whole brain eddy current distortions than residual in-plane distortions, especially for appropriate intersubject comparison of DTI data (Mohammadi et al., 2010). Especially if high b-values are used, special procedures should be employed to yield a set of coregistered template images wherein each image can be used as a reference image for the registration of a matching high b-value image (Ben-Amitay et al., 2012).

2.4 Motion Artifacts

Any MRI sequence is obviously sensitive to motion, but diffusion MRI is particularly sensitive to the presence of motion.

> Why?

Diffusion imaging is practically based on creating sufficient incoherent phase shifts to produce signal decay due to microscopic thermal motion. This is done by applying long and strong diffusion-weighting gradients, which unfortunately, is its greatest challenge. The reason is that the presence of any kind of macroscopic motion in the direction of the applied diffusion-weighting gradient may result in changes of voxel phase, which in turn results in a loss of voxel magnitude from spin phase dispersion, reflecting the range of motion within each voxel.

The longer and stronger the gradient pulses, the more direction changes of the molecules, and hence the bigger the loss of coherence and signal attenuation because of the macroscopic motion. Obviously, this macroscopic motion will be random (e.g., head motion) and in that sense, it cannot be identically repeated and dealt with. The problem gets even larger when considering that the sequence is repeated several times during an acquisition and obviously the motion will always be different. Hence, a distribution of different phase shifts will be appointed to different echoes, resulting in the well-known ghost artifacts along the phase-encoding direction. It has to be mentioned here, that the sensitivity in motion differs among different diffusion imaging methods, but this is out of the scope of this chapter. For the interested reader, details can be found elsewhere (Norris, 2001).

In Figure 2.3, typical motion artifacts, artifact-free DWIs, and B0 images from the work of Li et al. (2014) are shown. Figure 2.3a demonstrates complete signal loss caused by sudden head motion with large amplitude. Figure 2.3b demonstrates local signal loss, and Figure 2.3c the mismatching between slices due to involuntary motion.

Pulsatile motion is also a problem in diffusion imaging since the velocity of the fluids in the human body is usually uneven in different positions of the imaging plane. This means that, as mentioned earlier, the dephasing due to motion cannot be distinguished from dephasing due to microscopic diffusion, resulting in less prominent artifacts compared to rigid body motion. One

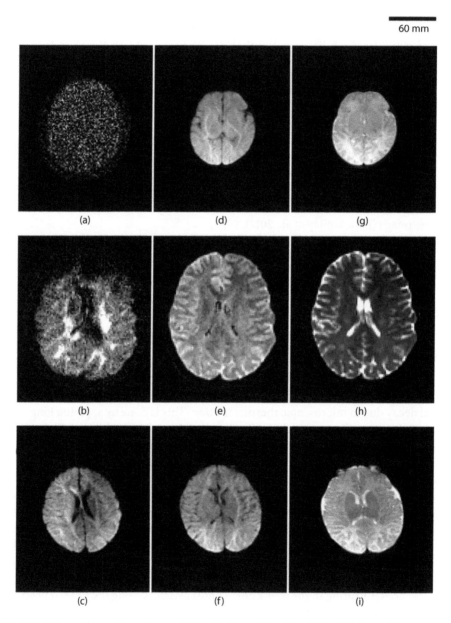

FIGURE 2.3 Illustrations of motion artifacts (a-c), artifact-free DWIs (d-f), and B0 images (g-i). Motion artifacts include complete signal loss (a), local signal loss (b), and mismatching (c). (From Li, X. et al., *PLOS One, 9*(4), *e94592*, 2014.)

typical example of this artifact is the pseudo-increase of the diffusion coefficient on the lateral ventricles because of the pulsatile motion of the cerebrospinal fluid (CSF), which makes isotropic diffusion in the ventricles to appear anisotropic (Koch and Norris, 2010). Moreover, at the interface between brain parenchyma and CSF, artifacts in the calculation of ADC from multiple b-values may arise due to partial volume effects. These effects appear because in the presence of motion, a given voxel is not at the same exact place throughout the different b-value dataset, leading to a slight overestimation of the brain ADC when compared to CSF ADC. It is calculated that the maximum displacement of the brain parenchyma originating from brain pulsation (due to cardiac cycle) is about 0.5 mm. It follows that the local misregistration between the actual image and DWI will be more severe at the boundaries of different tissue/air/bone properties.

2.4.1 Motion Artifacts—Mitigating Strategies

There are several ways to cure motion artifacts, depending on the specific cause. To mitigate artifacts from pulsatile motion, cardiac gating can be used in order to compensate for signal loss or ghosting. Unfortunately, cardiac gating can be demanding since it can increase the acquisition time and is not considered always reliable. Nevertheless, other researchers argue that if correctly implemented, cardiac gating can be performed in a very time-efficient way with very little increase in total scan time (Pierpaoli, 2011).

Regarding bulk motion, the only absolute way to eliminate it would be to use anesthesia (not very practical of course!) or use sequences so fast that motion would be effectively frozen during the acquisition. The most widely used sequence in diffusion imaging, due to its speed as well as high signal to noise ratio (SNR) is Echo Planar Imaging or EPI (Mansfield, 1977). EPI is now considered the gold standard in clinical routine diffusion MRI since an image is acquired within a single shot with a typical duration of about 100 ms, hence motion can be considered effectively frozen during the acquisition (Le Bihan et al., 2006).

The partial volume effects mentioned earlier can still appear even with the use of EPI. Nevertheless, they can be remedied by using inversion recovery prepared sequences instead of plain DWI in order to suppress the CSF signal diminishing the effect (Andersson et al., 2002; Fries et al., 2009).

Especially regarding high b-value diffusion MR imaging, which provides enhanced contrast compared to "regular" diffusion MRI, motion is a definite drawback and has to be dealt with. The most frequently adopted approach is to use image registration techniques to coregister each DW image to the first image collected (the B0) using it as a reference image. According to Ben Amitay et al. (2012), given the orientational sensitivity of the DW-MRI signal, cost functions such as cross correlation and least-squares (used for intramodal coregistration) are ineffective while cost functions such as mutual information (Studholme et al., 1999) or its normalized variant (Haselgrove and Moore, 1996) are used successfully.

2.4.2 EPI Specific Artifacts

EPI is the most widely used sequence in diffusion imaging due to its speed (the read-out train length of single-shot EPI is less than 100 ms), and its high SNR. However, although EPI was introduced as the cure for diffusion motion artifacts, it was proven that often a problem's cure may result in further problems.

Unfortunately, EPI was no exception since it is very vulnerable to artifacts and more specifically (but not limited to): (1) distortions originating from B0 inhomogeneities, (2) misregistration artifacts from Eddy currents, and (3) subject motion.

2.4.3 Distortions Originating from B0 Inhomogeneities

Figure 2.4 shows typical EPI geometric distortion artifact obtained by purposely mis-setting the x_2–y_2 shim value when acquiring data from a phantom in diffusion MRI (Reproduced from In et al., 2015). It is evident that EPI is sensitive to any local field offset, which may lead to artifacts in the pixel's apparent location in the reconstructed image. Fortunately, this suggests that the EPI geometric distortion problem is a one-dimensional problem of local warps amounted in the phase encode direction. Hence, if the particular field inhomogeneity is known, then it is possible to correct for these warps, so-called "unwarping."

Please note that there is a notion that EPI distortions do not produce major artifacts on calculated diffusion data because if distorted by B0 inhomogeneity, all images are equally distorted and therefore there is no need for correction prior to DTI analysis. Unfortunately, this does not apply for cases of severe distortions, and moreover, the diffusion data may be correct (in the sense that they will be scaled by the same factor), but will be anatomically incorrect, precluding image fusion and direct anatomical matching (Jezzard, 2012).

Hence, EPI distortions should always be corrected for, but extra data such as additional images for B0 mapping or structural images for target registration are needed (Bastin, 2001; Bodammer, et al., 2004). However, some groups have proposed techniques where no additional images are needed (Andersson and Skare, 2002; Andersson et al., 2008; Papadakis et al., 2005).

2.4.4 Misregistration Artifacts from Eddy Currents and Subject Motion

The two main mechanisms by which Eddy currents may produce corruption of diffusion measurements are

1. *Image distortions*
2. *Unwanted contributions to the b-value (i.e., the strength and duration of the gradient pulse)*

Data post-processing can reliably correct for image distortions in the clinical setting, but the quantification and correction of the unwanted diffusion sensitization produced by eddy currents is impractical (Andersson and Skare 2002; Jones et al., 2013). Hence, it is crucial to minimize

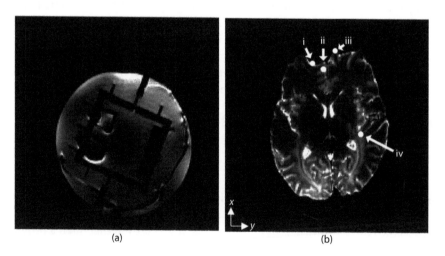

FIGURE 2.4   Distorted SE-EPI images of the phantom (a) and human brain (b) obtained by purposely mis-setting the x_2–y_2 shim value in diffusion MRI. (Modified from In, M. et al., *PLOS One 10*(2), e0116320, 2015.)

eddy currents before the acquisition by using optimal hardware design such as additional wiring or reduction of conductive surfaces, pre-emphasis of the gradients with proper calibration (Papadakis et al., 2000 ; Schmithorst and Dardzinski, 2002), and modified diffusion experiments that are less likely to produce eddy current artifacts, such as alternating polarity (Reese et al., 2003).

Subject motion is another source of artifacts in EPI as in any MRI data. The main consequence of motion is severe ghosting in the images especially in multishot acquisitions. Fortunately though, as mentioned earlier, clinical diffusion images, are acquired with single-shot EPI sequences rather than multishot acquisitions, and these are much less sensitive to image degradation due to motion.

2.4.5 Mitigating Strategies—EPI Specific

As already mentioned, there can be two approaches regarding EPI distortion corrections. One approach would be to correct the data once they have been collected and deal with the distortions they produce (Andersson and Skare, 2002; Papadakis et al., 2005). The other approach would be to try to avoid these distortions before their production, by altering the data collection techniques, such as using parallel imaging to minimize the readout time or modification of the diffusion experiment with bipolar diffusion gradients (see previous section) (Papadakis et al., 2000; Reese et al., 2003).

It seems that neither approach is better than the other. Once again it is verified that "many hands make light work." Hence theoretically, the highest accuracy and reliability of data collection should be pursued, and if considerable distortions persist, post-processing approaches should be used to mitigate them.

Figure 2.5 shows an example of advanced EPI distortion correction methods. The reference image is without distortion correction (A), the corresponding distortion-corrected fractional anisotropy (FA) maps with the extended point spread function (PSF) method with a reversed

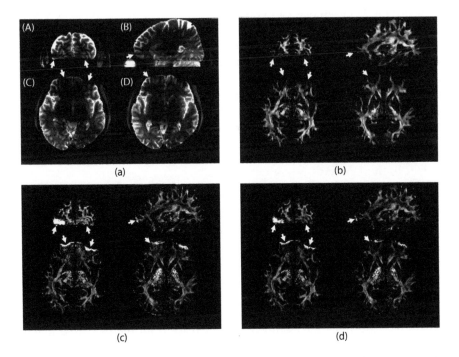

FIGURE 2.5 Example of advanced EPI distortion correction methods. A reference image (a) without distortion correction, the corresponding final distortion-corrected FA maps (b), the original reversed gradient approaches using algorithms of Jacobian modulation (c) and least-squares restoration (d). (Adapted from In, M. et al., *PLOS One, 10*(2), e0116320, 2015.)

phase gradient approach (B), the original reversed gradient approaches using algorithms of Jacobian modulation (C), and least-squares restoration (D). For comparison, a coronal (Figure 2.5a), sagittal (Figure 2.5b) and two axial images Figures 2.5c and d were selected, and white arrows point to the differences between FA maps (In et al., 2015).

The arrows are marked at the same locations in all images to facilitate appreciating the degree of local distortion. It is evident that post-processing with all different distortion correction procedures has successfully reduced the distortion (Figures 2.5c and d), although improved correction with the extended PSF method (Figure 2.5b) can be found, especially in regions with strong geometric distortions (In et al., 2015).

2.5 Artifacts Due to Properties of the Subject Being Imaged and "Physiological" Noise

2.5.1 Susceptibility-Induced Distortions

The human body can induce magnetic field alterations due to its inhomogeneous consistency. These susceptibility-induced distortions have a particularly severe effect on DWI, producing degradation of images especially with EPI sequences. The reason is that EPI sequences require a very homogeneous magnetic field, but the truth is that there is no such thing!

> Why?

Because even in a perfectly homogeneous magnetic field of a well-shimmed magnet, the introduction of a human head (with air cavities, bone, fat, etc.), would certainly locally distort it. These discontinuities of magnetic susceptibility, induced by field inhomogeneities around the boundaries of air and tissue, produce local magnetic field gradients that will interact with the diffusion gradients. These additional local gradients can then cause spatial variations of the b-matrix resulting in image distortion or signal drop-out (Basser and Jones, 2002). In neuroimaging in particular, this is a typical element of concern since we usually need to examine fused images (e.g., white matter tracts superimposed on higher resolution structural images) and these might not register properly, revealing a significant pitfall.

2.5.2 Physiological Noise

Lately, the term "physiological noise" is increasingly discussed and used in the DWI community, although this term was first used and for many years constrained within the fMRI community. It obviously refers to signal variability, which is related to involuntary motion of the patient such as cardiac or respiratory motion (Friston et al., 1995; Marchini and Smith, 2003).

In a broader sense, physiological noise may include all sources of data variability excluding the so-called uncorrelated noise (that is, quantum and thermal noise or system instabilities), that may be related to scanning a living subject that breathes and moves. Unfortunately, especially with the increase of magnetic fields from 1.5T to 3T and recently to 7T, physiological noise tends to increase since the assumed SNR increase from the increased magnetic field only applies to thermal noise (Triantafyllou et al., 2005).

Physiological noise has been extensively studied for fMRI but not as much for DWI experiments. Nevertheless, the increase of clinical diffusion studies with higher magnetic fields, as well as the demands for accuracy and precision in diffusion measurements, is expected to multiply the investigations and analyses of the potential effects of physiological noise. Already there is a number of recent studies (Chang et al., 2012; Kreilkamp et al., 2016; Mohammadi et al., 2010; Walker et al., 2011) that sought to investigate and characterize the potential effect of physiological noise artifacts, especially those originating from cardiac pulsation and subject motion, and propose methods to effectively remove them. Moreover, a few studies have investigated the potential errors on the statistical analysis of diffusion tensor MRI results from physiological noise (Peterson et al., 2008; Walker et al., 2011). These will be also discussed in Section 2.7

2.5.3 Susceptibility Effects and Physiological Noise—Mitigating Strategies

DWI suffers from susceptibility effects and, especially on higher field strength scanners (>3T), which are increasingly used in neuroradiology, these problems become more extensive.

Generally, there are two approaches regarding the correction of geometric distortions induced by magnetic susceptibility effects.

One way is to use a field mapping (a scaled version of the B0 map) to correct for every single voxel displacement with theoretical models of how field inhomogeneities distort the images (Jezzard et al., 1998, 2012). In the majority of such correction methods, two consecutive steps are involved that differ in the way the inhomogeneous field map is measured. Either the difference between the assumed and actual magnetic field, ΔB, is measured, or the effect of ΔB is accounted for in the process of image reconstruction so that an undistorted, or spatially corrected, image is produced (Morgan et al., 2004).

The second approach is called the "reverse gradient" method and was first proposed by Chang and Fitzpatrick in 1992. This method uses two different EPI acquisitions, one from "bottom up," and a second from the other way around, so that a pair of "reverse" images is produced. The distortions in those images are theoretically of the same magnitude but opposite in direction. Hence, the signal can be recovered provided that the displacements due to field perturbations can be well described by a theoretical model.

Regarding physiological noise, and especially for DTI data, the approaches for reducing artifacts are (1) cardiac gating and (2) robust tensor estimation.

Cardiac gating was originally proposed to reduce the incidence of the cardiac pulsation misplacement artifacts in the early 1990s (Turner et al., 1990). However, the clinical application of cardiac gating is still very restricted due to the reported increase in scan time, since it is true that the availability of time efficient cardiac gating is limited in commercial scanners. Nevertheless, more and more recent studies indicate that cardiac gating is incomparable in reducing the effects of cardiac pulsation and propose ways to reduce the time penalty using triggering and sampling variations (Gui et al., 2008; Kim et al., 2010; Nunes et al., 2005).

On the other hand, robust tensor estimation is a technique that allows the identification of data points originating from artifacts and their removal from the contribution to the diffusion tensor estimation. This technique is called "outlier rejection," and it has been shown that these outliers are more often presented in the cerebellum, CSF spaces, brainstem, etc., and in general in regions that are more affected by cardiac pulsation (Peterson et al., 2008). Nevertheless, in the algorithm by Chang et al. (2005), called RESTORE (**R**obust **ES**timation of **T**ensors by **O**utlier **RE**jection), it is evident that these kinds of methods can be very effective in dealing with other artifacts as well, not only those originating from cardiac pulsation, especially for automated data-analysis procedures (Chang et al., 2005).

Finally, the **P**eriodically **R**otated **O**verlapping **P**arall**EL** **L**ines with Enhanced **R**econstruction method, otherwise known as PROPELLER, an EPI sequence can be used instead of plain EPI.

This sequence is designed to collect data in strips, also known as blades (Pipe et al., 2002), with every blade measuring data in the k-space center. After data collection, all blades can be combined to produce the image, which contains all the necessary information (low spatial frequency) to characterize any corruptions from motion. In the study from Pipe and Zwart (2006), a revised version of PROPELLER called Turboprop is described for improved DWI, resulting in enhanced diffusion images and making post-acquisition correction methods more robust.

2.6 Processing and Interpretation Pitfalls

This section is dedicated to the processing and interpretation pitfalls investigators may encounter during the analysis of diffusion data, which can be thought of more as limitations of the methods and techniques than artifacts resulting from the MRI scanner. In that sense, we need to examine several steps of the analysis pipeline, which can be grouped into the following three distinct categories:

1. Preprocessing of data
2. Quantitation of parameters
3. Intrasubject and intersubject comparisons

It should be noted at this point that this section is not going to cover in depth all the involved potential pitfalls; we are referring to procedures with multiple steps and each step can be susceptible to many sources of bias, creating a rather large field that cannot be covered in one section. Nevertheless, this will be an introduction to the fundamental potential errors involved, which if addressed correctly, can ensure reliable data and meaningful results. Furthermore, this section has been largely based on an excellent recent review paper by Jones and Cercignani (2010) where the aforementioned analysis pipeline is fully examined, hence the interested reader should refer to that paper for more specific information and details.

2.6.1 Preprocessing of Data

We have already discussed in detail the susceptibility effects as well as their mitigating strategies in the previous sections. The most popular method for correcting these susceptibility-induced distortions is so-called "field mapping," which allows the local shift (induced by the field inhomogeneities that contribute as a "background gradient") to be calculated and corrected on a voxel-by-voxel basis.

The problem is that it might be possible that signal intensities from neighboring voxels may contribute to the voxel of measurement, leading to the so-called "many-to-one-mapping" pitfall. That is, after unwarping, signals from several voxels may collapse into a single voxel (due to field inhomogeneity) and then these signals are incorrectly "demodulated," resulting in identical tensors computed and assigned to more than one region, appearing as structures that are inconsistent with the known brain anatomy (Jones and Cercignani, 2010).

An example of this pitfall is shown in Figure 2.6. Using the standard field mapping approach, FA images obtained before and after unwarping, for the same subject and the same slice location, are depicted. The T2-weighted image of the same slice can be used for comparison. The problem is that after the unwarping, two structures that are inconsistent with any known anatomy appear (two perfectly straight fibers with uniform anisotropy as indicated by the arrows) and are therefore a pitfall.

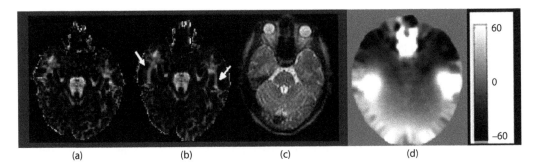

FIGURE 2.6 FA maps computed from data before (a) and after (b) unwarping using the standard field mapping approach. A T2-weighted anatomical image (c) is also shown for comparison. The corresponding B0 map in Hertz is shown in (d). (Adapted from Jones, D. K. and Cercignani, M., *NMR Biomed.*, 23, 803–820, 2010. With permission.)

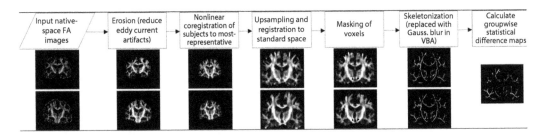

FIGURE 2.7 Flowchart of the steps of the original TBSS procedure. Under CC BY 3.0 license. (From Schwarz, C. G. et al. *NeuroImage*, 94, 65–78, 2014.)

The "many-to-one-mapping" problem can be better dealt with if instead of using the field mapping approach one uses the reverse gradient method, since with this method the orientational information is better preserved. Although the time of the calculation is going to be doubled as twice as many data are required.

It is evident that correct anatomical alignment in neuroimaging studies is of paramount importance, and considerable effort is put into providing robust registration to improve spatial correspondence. (Andersson and Sotiropoulos, 2016; de Groot et al., 2013; Jbabdi et al., 2012).

One of the most popular methods to coregister sets of diffusion tensor FA images for performing voxelwise comparisons is called tract-based spatial statistics (TBSS) (Smith et al., 2006). It establishes spatial correspondence using a combination of nonlinear registration and a so-called "skeleton projection" that may break the topological consistency of the transformed brain images.

The reason is that the alignment of FA images from multiple subjects with voxelwise analysis does not yet allow the extraction of valid conclusions, especially when the choice of spatial smoothing remains arbitrary. Hence, TBSS aimed to solve the aforementioned issues by a carefully tuned nonlinear registration, followed by the so-called "mean FA skeleton," which is a projection onto an alignment-invariant tract representation. This way, it is anticipated that the sensitivity, objectivity, and interpretability of analysis of multisubject diffusion imaging studies will improve (Smith et al., 2006).

A flowchart with the steps of the original TBSS procedure is depicted in Figure 2.7.

Nevertheless, the improvement of symmetry is not always absolutely correct, hence a careful inspection of the results is always recommended. Moreover, it has recently been suggested that improved DTI registration allows voxel-based analysis that may outperform TBSS (Schwarz et al., 2014).

2.6.2 Quantitation of Parameters

Other pitfalls are specific to the quantitation of DTI parameters due to biases in the calculation of the eigenvalues (and associated eigenvectors) or to the limitation of the algorithms used for tracking. Especially regarding DTI tractography, the reconstruction of fiber bundles completely depends on the quality of the estimated eigenvector. The pathway is built based on the direction of maximum diffusion, so it will obviously fail if the information is interrupted or produce false results if the information is corrupted from acquisition noise. A schematic representation of how a hypothetical error on the direction of diffusivity in a certain voxel can affect the overall result is depicted in Figure 2.8.

Nonetheless, the limitation with the greater impact in fiber tractography is the fact that most models assume homogeneous distribution of fibers inside a voxel following a free Gaussian distribution. Unfortunately, this is not the case. In reality, voxels contain several distributions of fibers that may merge, split, bend, or cross. Hence, the overall principal eigenvector that a voxel will yield using those algorithms may be far from correct, giving an erroneous result regarding the underlying structure orientation. A simple but very indicative example of this problem is illustrated in Figure 2.9. Two fibers that are crossing at 90° would yield a tensor that is far from the true orientation and may be the source of several problems in tractography, such as interruption of the tract or failed connections between neighboring fibers.

Several models have been proposed to overcome this limitation of the DTI procedure, with the most popular being the high angular resolution diffusion imaging (HARDI) group of models, which try to mitigate the problems that arise from the assumption of Gaussian distribution. This is obviously done at a cost of huge calculation times due to the excessive amount of data needed for the model to be as accurate as possible.

In fact, early in the development of DTI based tractography, a flaw in the diffusion tensor model was quickly noticed. This was the assumption that there is a single ellipsoid in each imaging voxel as if all the axons traveling through a voxel traveled in exactly the same direction (Tuch et al., 2002). It has been proposed that in more than 30% of the voxels in a standard resolution brain image, there will be at least two different neural tracts traveling in different directions that pass through each other. In the classic diffusion ellipsoid tensor model, the information from the crossing tract just appears as noise or unexplained decreased anisotropy in a given voxel.

It would be easier to grasp the idea by conceptually placing a kind of geodesic dome around each image voxel. Initially this was described as an "icosahedron," which provided a mathematical basis for passing a large number of evenly spaced gradient trajectories through the voxel, each coinciding with one of the apices of the icosahedron.

In geometry, an icosahedron is a polyhedron with 20 faces. The name comes from the Greek είκοσι (*eíkosi*), meaning "twenty," and έδρα (*hédra*), meaning "seat."

More specifically, a large number of directions (>40) are currently being used by adding more evenly spaced apices to the original icosahedron (Hext, 1963). Nevertheless, the gradient strengths have to be considerably higher than for standard DTI since, as already mentioned, the apparent noise (nondiffusion contributions to signal) must be reduced using higher b-values and improving the spatial resolution.

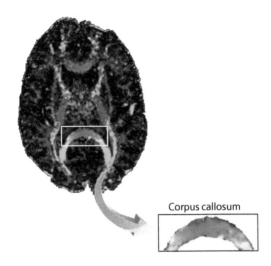

Corpus callosum

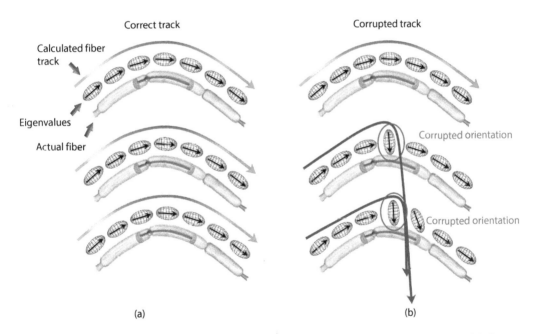

Correct track   Corrupted track

Calculated fiber track

Eigenvalues

Actual fiber

Corrupted orientation

Corrupted orientation

(a)   (b)

FIGURE 2.8 A schematic representation of how a hypothetical error on the direction of diffusivity in a certain voxel can affect the tractography results. (a) Correctly calculated fiber tracks in the corpus callosum. (b) Scheme showing how a fiber track can be erroneously estimated by a corrupted orientation (for instance by acquisition noise or geometric distortion).

It has also been proposed that "very high angular resolution" (>256 directions) can be used, but this would result to 2000 or more images, which obviously makes such analysis practically impossible.

Hence, a further implementation of the HARDI approach is the Q-Ball method, which is a mathematical alternative to the tensor model. It was introduced by David Tuch, where instead of forcing the diffusion anisotropy data into a group of tensors, probability distributions, geometric tomography, and vector math were used (Tuch, 2004). This is illustrated in Figure 2.10,

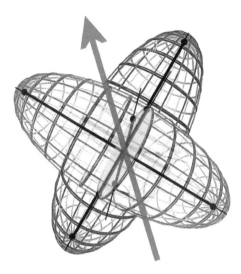

FIGURE 2.9 Two fibers that are crossing at 90° would yield a tensor (arrow) that might be far from the true orientation, causing track interruption or failed connections between neighboring fibers.

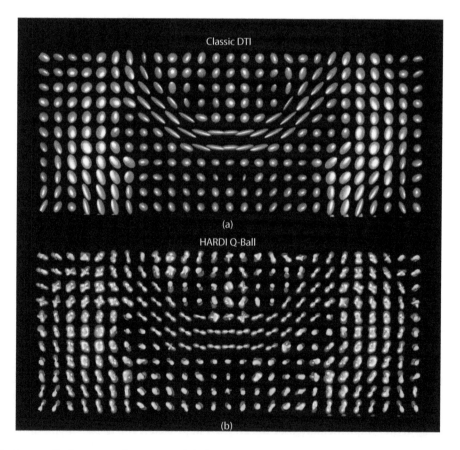

FIGURE 2.10 Difference in the intravoxel diffusion pattern between the simpler classic DTI (a) and its successor, the HARDI Q-Ball technique (b). (Courtesy of Vesna Prckovska.)

where it is evident that the HARDI Q-Ball technique is able to better capture the intravoxel diffusion pattern compared to its simpler predecessor, the DTI approach.

2.6.3 Dependence of Estimated Mean Diffusivity on b-Factor

Eigenvalues and the associated eigenvectors are considered rotationally invariant parameters. Nevertheless, they are subject to several biases, and in that sense they can affect the scalar indices used in the quantification of diffusion results, namely, the mean diffusivity and the FA.

The dependence of the estimated mean diffusivity (trace) on the b-factor is a pitfall that needs to be considered, especially in comparing results among different multicenter studies, where obviously the same b-value should be used.

> But if the trace is indeed the mean square displacement of water molecules per unit time (multiplied three times), then why it is affected by the strength and duration of the gradient pulses?

Again, we need to remember that we assume that the displacement of water molecules (in three dimensions) follows a Gaussian distribution even though this is not true. In practice, water molecules in the brain do not generally follow a Gaussian distribution, and the b-value plays an important role in signal attenuation from diffusion. Yoshiura et al. (2001) showed that as the diffusion weighting is increased, the dissociation between white and gray matter increases due to nonmonoexponential diffusion signal decay of the brain tissue, which is more prominent in white matter than in gray matter. This is shown in Figure 2.11, where the variation of the mean diffusivity as a function of b-value is depicted.

As shown by Pierpaoli et al. (2011), at b-values of the order of 1000 smm^{-2}, the trace is remarkably uniform throughout parenchyma, but the problem is that there is no specific agreement among centers on what the b-value should be used for diffusion tensor experiments. The take-home message should be that in a certain measurement within a region of interest (ROI), the mean and variance of the trace will be dependent on the b-value and will differ among centers. This means that comparison with normative databases or data from other centers and, therefore, multicenter studies, are going to be problematic unless the same degree of diffusion-weighting is employed (Jones and Cercignani, 2010).

2.6.4 Effect on ROI Positioning and Bias on Parametric Maps

The simplest approach for the extraction of parameters for the evaluation of a certain brain region is the use of the ROI measurements. Typically, a ROI is manually placed on the image to measure the indices of interest. Here, the problem is the selection of the most suitable image for the ROI placement.

> Why?

Because placing the ROI directly on the anisotropy image proves to be inaccurate, hence a more suitable anatomical image is needed. Therefore, a "background" image is used, such as the corresponding T2 or T1 weighted images placed behind the anisotropy image. However, this depends on the quality of the matching of the two images due to geometric distortion as

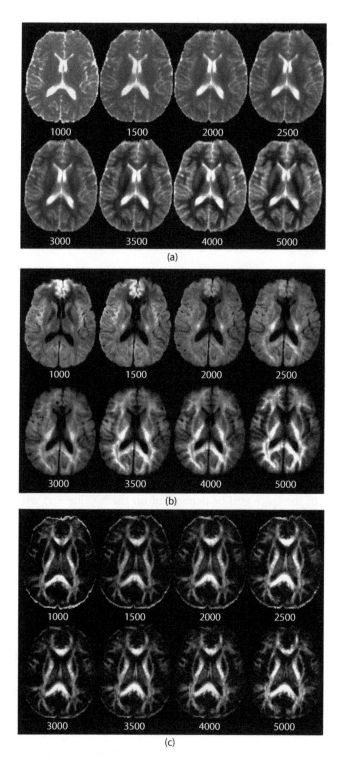

FIGURE 2.11 Trace ADC maps (a) and isotropic DW images (b) as a function of b-value. The number below each image represents the b-value in s/mm². Note changes of gray–white contrast in relation to b-value in the trace ADC maps and isotropic DWIs. (Adapted from Yoshiura, T. et al., *Magn. Reson. Med.*, 45(5), 734–740, 2001. With permission.)

previously discussed (see EPI specific distortions). A "safe" choice is to always use the b = 0 image where the contrast is independent of diffusion.

Nevertheless, even after image realignment techniques, registration errors might persist and this is more prominent when positioning small ROIs in subcortical regions since even a small ROI misplacement can have a large effect on FA measurement due to the large variations of anisotropy between neighboring tissues.

2.6.5 CSF Contamination in Tract Specific Measurements

An alternative way to conduct ROI studies in particular fiber tracts is to first determine the tract by using a fiber-tracking algorithm, and then extract the mean parameters for evaluation. Several studies report that this approach obviously increases the sensitivity and specificity of the particular measurement since the delineation is not performed by hand; however, it is not free from errors that may appear due to partial volume effects (Jones et al., 2005; Kanaan et al., 2006; Shergill et al., 2007).

Jones et al. (2005), described a method to check for the aforementioned error by using color-encoding of mean diffusivity directly on the reconstructed tract so that any partial volume effect will be immediately apparent, especially between fairly inhomogeneous regions of the brain, such as between parenchyma and CSF. Moreover, the use of CSF suppression during the acquisition, such the use of **FL**uid-**A**ttenuated **I**nversion **R**ecovery diffusion sequences (FLAIR DWI) can mitigate such artifacts, substantially improving fiber tracking results (Chou et al., 2005). On the other hand, Pasternak et al. (2009), proposed a method that obtains enhanced diffusion indices for the case of partial volume effects with free water. It included the use of so-called "bi-tensor separation" (one tensor model for the free water and one for the tissue) together with additional biological and physical constraints, and it was demonstrated that the contribution of the free water compartment can be differentiated from that of the tissue compartment. They suggested that **F**ree **W**ater **E**limination (FWE) can be performed as preprocessing for any DTI-related analysis, both in healthy and clinical cases.

Last, in a very recent study, Hoy et al. (2015) evaluated the ability of FWE DTI methods to minimize the partial volume effects of CSF for deterministic tractography applications, and suggested that the FWE diffusion model overcomes CSF partial volume effects, without the time, SNR, and volumetric coverage penalties inherent to FLAIR DTI.

2.6.6 Intrasubject and Intersubject Comparisons

Most groups that use diffusion tensor studies for clinical research use the following approach to calculate image comparisons and performance variables across subject groups: they first try to coregister all subjects and then perform statistical tests for groupwise differences in each coregistered voxel, a method known in analysis of structural MRI as voxel-based analysis (VBA) or voxel-based morphometry (VBM).

VBM was originally designed to measure changes of gray matter on structural T1-weighted data; nevertheless it is becoming more and more popular as a method for analyzing quantitative images since it is quite automated with minimal intervention from the user.

Unfortunately, VBM can be highly sensitive to registration errors and may produce false positives in affected regions (Ashburner and Friston, 2001; Bookstein, 2001).

As mentioned already (see section 2.6.1) Smith et al. introduced TBSS, which attempts to reduce the effects of local misregistrations by projecting all FA voxels onto the nearest location on a "skeleton," approximating white matter (WM) tract centers (Schwarz et al., 2014;

Smith et al., 2006). TBSS has been widely used through the very popular Oxford Centre for Functional MRI of the Brain (FMRIB) Software Library (FSL) (Jenkinson et al., 2012), although several groups continue to use standard VBM, alone or in combination with TBSS (Chiang et al., 2011; Douaud et al., 2011).

2.7 Mitigating Strategies—Available Methods and Software for Diffusion Data Correction

A variety of software packages for the preprocessing and analysis of diffusion data are available, including (but not limited to) tensor estimation, tractography, ROI analysis, registration, etc. In the following section, a brief informative description of the most commonly used packages is included so that the interested reader can proceed to a more comprehensive analysis, evaluation, and comparison. The majority of these software packages with their basic characteristics are depicted in Table 2.1.

2.7.1 RESTORE Algorithm

The RESTORE algorithm is an effective method that improves tensor estimation on a voxel by voxel basis (Chang et al., 2012). Namely, it utilizes weighted nonlinear least-square tensor fitting methods in order to detect and remove artifactual data points (outliers) in DWIs. In this

TABLE 2.1   DWI/DTI Analysis Software Packages

Software Name	Preprocessing	Tensor Estimation	Tractography	ROI Analysis	Registration
AFNI	✓	✓			
BioImage Suite		✓	✓	✓	
BrainVoyager QX		✓	✓		
Camino		✓	✓		
Dipy		✓	✓		
DoDTI	✓	✓	✓		
DTI-Query			✓		
DTI-TK					✓
DTIStudio		✓	✓	✓	
ExploreDTI	✓	✓	✓		
FreeSurfer	✓	✓			
FSL-FDT	✓	✓	✓		
FSL-TBSS					✓
JIST	✓	✓			
MedINRIA		✓	✓	✓	
MrDiffusion		✓	✓		
MRtrix		✓	✓		
SATURN		✓	✓		
SPM	✓	✓			
TrackVis		✓	✓	✓	
TORTOISE	✓	✓		✓	
3D Slicer		✓	✓	✓	

robust estimator approach is the tensor, which is first estimated using all the data measurements, and subsequently, the data points with differences between calculated and measured DW values, are rejected.

As previously discussed, there is an number of artifacts, which need to be addressed. The RESTORE method is applied to reduce the contribution of artifacts, originated from subject motion and cardiac pulsation (Chang et al., 2005). Brain regions known to be affected by cardiac pulsation are the cerebellum and the genu of the corpus callosum (Walker et al., 2011). Due to the voxel by voxel processing, this algorithm can exclude the voxels that contain artifactual data without affecting voxels with no artefactual signals in a partially corrupted image. RESTORE is widely implemented as a part of the workflow in DTI analysis software packages such as *ExploreDTI*.

2.7.2 ExploreDTI

ExploreDTI is a MATLAB®-based graphical toolbox developed for DTI analysis and fiber tractography. More specifically, it is focused on interactive display and manipulation of data (such as white matter fiber tracks), brain surface renderings, and diffusion glyphs (Leemans et al., 2009).

The first step of the workflow is the input of datasets, which can be in one of the acceptable formats (Dicom, NIFTI, or MATLAB). During the processing step, a linear, nonlinear, and weighted linear least-square method is implemented, with or without the RESTORE algorithm, in order to estimate the diffusion tensor. Concerning tractography, ExploreDTI uses deterministic and wild bootstrap streamline tracking algorithms and offers a wide range of options, such as FA calculation, relative anisotropy, mean diffusivity, and other modalities (Leemans et al., 2009). Subject motion and eddy current-induced distortions are diffusion artifacts that can be effectively treated with ExploreDTI in the preprocessing stage. Weblink: http://www.exploredti.com/.

2.7.3 FSL-FDT

The FSL is a comprehensive library of analysis tools for MRI, DTI, and fMRI. FDT, as a part of FSL, is a software tool for analysis of DW images and offers a variety of tools for data preprocessing, local diffusion modeling, and tractography.

There are four main FSL-FDT programs, and one of them regards artifact correction in DTI. *"Probtrackx"* is a component program for probabilistic tractography and connectivity-based segmentation, and *"Bedpost"* is a program for local modeling of diffusion parameters, especially crossing fibers and uncertainty. However, *"eddycorrect"* should be run first in order to correct the eddy current distortions, which are different for different gradient directions. At the end, *"dtifit"* estimates a diffusion tensor at each voxel of the brain and outputs a set of FA images, mean diffusivity (MD), tensor eigenvalues, and eigenvectors. A registration option is also available via **FMRIB's Linear Image Registration Tool** (FLIRT), the image registration tool of FSL. Weblink: http://fsl.fmrib.ox.ac.uk/fsl/fslwiki/FDT.

2.7.4 FreeSurfer—TRACULA

FreeSurfer is a set of tools for analysis and visualization of structural and functional brain MRI data, applying surface-based analysis. One of these tools is TRACULA (**TRA**cts **C**onstrained by **U**nder**L**ying **A**natomy), a method for automated reconstruction of major white matter pathways that is based on the global probabilistic tractography with anatomical priors (Yendiki et al., 2011).

A typical workflow analysis begins with the input of a T1-weighted MR image and a DW image. Mitigation of both eddy current distortions and subject motion is achieved by registering the B0 image to the T1-weighted MRI scan and to a common atlas (Yendiki et al., 2011). Subsequently, using the T1 image data, FreeSurfer performs a subcortical segmentation and a cortical parcellation. The aforementioned atlas includes 18 white matter pathways; therefore, the reconstruction of volumetric distribution is atlas-based. The final output of the whole procedure is a file with exact statistics on many diffusion measures (e.g., FA or MD) for all reconstructed pathways. Weblink: https://surfer.nmr.mgh.harvard.edu/fswiki/Tracula.

2.7.5 TORTOISE

TORTOISE (**T**olerably **O**bsessive **R**egistration and **T**ensor **O**ptimization **I**ndolent **S**oftware **E**nsemble) is a comprehensive software package for both diffusion tensor estimation and diffusion artifact correction. It is a standalone application and consists of two parts, DIFF_PREP and DIFF_CALC. DIFF_PREP is the first module, where motion, eddy current distortion, and EPI distortion correction is performed. Namely, eddy current correction is usually conducted using an affine deformation model, while EPI distortion correction is done using registration to a structural MRI target without the need of B0 field mapping (Pierpaoli et al., 2010). DIFF_CALC, as the second module, regards tensor calculation and fitting with a weighted linear and nonlinear least-square method, error analysis, color map visualization, and ROI analysis. TORTOISE is commonly used as an artifact correction tool because of its ability to correct as a whole major diffusion artifacts, such as subject motion, eddy current, and EPI distortions. Weblink: http://www.tortoisedti.org.

2.8 Conclusion

As stated in the beginning of this chapter, the acquisition as well as the interpretation of diffusion imaging results is not always straightforward. The sensitivity of diffusion imaging experiments to phenomena other than diffusion and the technical difficulties of a quite long analysis pipeline, starting from acquisitions and hardware, to final evaluation and interpretation, can be indeed challenging.

However, the scope of this chapter was to introduce and analyze the artifacts and pitfalls in a qualitative way, and to evaluate possible mitigating strategies or solutions to overcome them. Hopefully this material can be a good starting point and a valuable reference in the difficult task of MR diffusion data evaluation and interpretation.

References

Andersson, L., Bolling, M., Wirestam, R., Holtas, S., and Stahlberg, F. (2002). Combined diffusion weighting and CSF suppression in functional MRI. *NMR in Biomedicine, 15,* 235–240.

Andersson, J. L. and Skare, S. (2002). A model-based method for retrospective correction of geometric distortions in diffusion-weighted EPI. *NeuroImage, 16*(1), 177–199. doi:10.1006/nimg.2001.1039

Andersson, J. L. and Sotiropoulos, S. N. (2016). An integrated approach to correction for off-resonance effects and subject movement in diffusion MR imaging. *NeuroImage, 125,* 1063–1078. doi:10.1016/j.neuroimage.2015.10.019

Andersson, J. L., Smith, S., and Jenkinson, M. (2008). FNIRT-FMRIB's non-linear image registration tool. *In 14th Annual Meeting of the Organisation for Human Brain Mapping*, Melbourne, Australia (pp. 496).

Ashburner, J. and Friston, K. J. (2001). Why voxel-based morphometry should be used. *NeuroImage, 14*(6), 1238–1243. doi:10.1006/nimg.2001.0961

Basser, P. J. and Jones, D. K. (2002). Diffusion-tensor MRI: Theory, experimental design and data analysis—A technical review. *NMR in Biomedicine, 15*(7–8), 456–467. doi:10.1002/nbm.783

Bastin, M. E. (1999). Correction of eddy current-induced artefacts in diffusion tensor imaging using iterative cross-correlation. *Magnetic Resonance Imaging, 17*, 1011–1024.

Bastin, M. E. (2001). On the use of the FLAIR technique to improve the correction of eddy current induced artefacts in MR diffusion tensor imaging. *Magnetic Resonance Imaging, 19*(7), 937–950. doi:10.1016/s0730-725x(01)00427-1

Bastin, M. E. and Armitage, P. A. (2000). On the use of water phantom images to calibrate and correct eddy current induced artefacts in MR diffusion tensor imaging. *Magnetic Resonance Imaging, 18*(6), 681–687.

Ben-Amitay, S., Jones, D. K., and Assaf, Y. (2012). Motion correction and registration of high b-value diffusion weighted images. *Magnetic Resonance in Medicine, 67*(6), 1694–1702.

Bodammer, N., Kaufmann, J., Kanowski, M., and Tempelmann, C. (2004). Eddy current correction in diffusion-weighted imaging using pairs of images acquired with opposite diffusion gradient polarity. *Magnetic Resonance in Medicine, 51*(1), 188–193. doi:10.1002/mrm.10690

Bookstein, F. L. (2001). "Voxel-based morphometry" should not be used with imperfectly registered images. *NeuroImage, 14*(6), 1454–1462. doi:10.1006/nimg.2001.0770

Chang, H. and Fitzpatrick, J. (1992). A technique for accurate magnetic resonance imaging in the presence of field inhomogeneities. *IEEE Transactions on Medical Imaging, 11*(3), 319–329. doi:10.1109/42.158935

Chang, L., Jones, D. K., and Pierpaoli, C. (2005). RESTORE: Robust estimation of tensors by outlier rejection. *Magnetic Resonance in Medicine, 53*(5), 1088–1095. doi:10.1002/mrm.20426

Chang, L., Walker, L., and Pierpaoli, C. (2012). Informed RESTORE: A method for robust estimation of diffusion tensor from low redundancy datasets in the presence of physiological noise artifacts. *Magnetic Resonance in Medicine, 68*(5), 1654–1663. doi:10.1002/mrm.24173

Chiang, M., Hickie, I. B., McMahon, K. L., Martin, N. G., Toga, A. W., Thompson, P. M. et al. (2011). Genetics of white matter development: A DTI study of 705 twins and their siblings aged 12 to 29. *NeuroImage, 54*(3), 2308–2317.

Chou, M. C., Lin, Y. R., Huang, T. Y., Wang, C. Y., Chung, H. W., Juan, C. J, and Chen, C. Y. (2005). FLAIR diffusion tensor MR tractography: Comparison of fiber tracking with conventional imaging. *AJNR American Journal of Neuroradiology, 26*, 591–597.

Douaud, G., Jbabdi, S., Behrens, T. E., Menke, R. A., Gass, A., Monsch, A. U., and Smith, S. (2011). DTI measures in crossing-fibre areas: Increased diffusion anisotropy reveals early white matter alteration in MCI and mild Alzheimers disease. *NeuroImage, 55*(3), 880–890. doi:10.1016/j.neuroimage.2010.12.008

Fries, P., Bücker, A., Kirchin, M. A., Naul, L. G., Runge, V. M., Reith, W., Stemmer, A. et al. (2009). Diffusion-weighted imaging in patients with acute brain ischemia at 3 T: Current possibilities and future perspectives comparing conventional echoplanar diffusion-weighted imaging and fast spin echo diffusion-weighted imaging sequences using BLADE (PROPELLER). *Investigative Radiology, 44*(6), 351–359.

Friston, K. J., Frackowiak, R. S., Grasby, P. J., Holmes, A. P., Poline, J. B., Turner, R., and Williams, S. C. (1995). Analysis of fMRI time-series revisited. *NeuroImage, 2*(1), 45–53.

Groot, M. D., Vernooij, M. W., Klein, S., Ikram, M. A., Vos, F. M., Smith, S. M. et al. (2013). Improving alignment in tract-based spatial statistics: Evaluation and optimization of image registration. *NeuroImage, 76*, 400–411. doi:10.1016/j.neuroimage.2013.03.015

Gui, M., Tamhane, A. A., and Arfanakis, K. (2008). Contribution of cardiac-induced brain pulsation to the noise of the diffusion tensor in Turboprop diffusion tensor imaging (DTI). *Journal of Magnetic Resonance Imaging, 27*(5), 1164–1168. doi:10.1002/jmri.21335

Haselgrove, J. C. and Moore, J. R. (1996). Correction for distortion of echo-planar images used to calculate the apparent diffusion coefficient. *Magnetic Resonance Medicine, 36,* 960–964.

Horsfield, M. A. (1999). Mapping eddy current induced fields for the correction of diffusion-weighted echo planar images. *Magnetic Resonance Imaging, 17,* 1335–1345.

Hoy, A. R., Kecskemeti, S. R., and Alexander, A. L. (2015). Free water elimination diffusion tractography: A comparison with conventional and fluid-attenuated inversion recovery, diffusion tensor imaging acquisitions. *Journal of Magnetic Resonance Imaging, 42*(6). doi:10.1002/jmri.25091

In, M., Posnansky, O., Beall, E. B., Lowe, M. J., and Speck, O. (2015). Distortion correction in EPI using an extended PSF method with a reversed phase gradient approach. *PLOS One, 10*(2), e0116320. doi:10.1371/journal.pone.0116320

Jbabdi, S., Sotiropoulos, S. N., Savio, A. M., Graña, M., and Behrens, T. E. (2012). Model-based analysis of multishell diffusion MR data for tractography: How to get over fitting problems. *Magnetic Resonance in Medicine, 68*(6), 1846–1855. doi:10.1002/mrm.24204

Jenkinson, M., Beckmann, C. F., Behrens, T. E. J., Woolrich, M. W., and Smith S. M. (2012). FSL. *NeuroImage, 62,* 782–790.

Jezzard, P. (2012). Correction of geometric distortion in fMRI data. *NeuroImage, 62*(2), 648–651. doi:10.1016/j.neuroimage.2011.09.010

Jezzard, P., Barnett, A. S., and Pierpaoli, C. (1998). Characterization of and correction for eddy current artifacts in echo planar diffusion imaging. *Magnetic Resonance in Medicine, 39,* 801–812.

Jones, D. K., Catani, M., Pierpaoli, C., Reeves, S. J., Shergill, S. S., Osullivan, M., and Howard, R. J. (2005). Age effects on diffusion tensor magnetic resonance imaging tractography measures of frontal cortex connections in schizophrenia. *Human Brain Mapping, 27*(3), 230–238. doi:10.1002/hbm.20179

Jones, D. K. and Cercignani M. (2010). Twenty-five pitfalls in the analysis of diffusion MRI data. *NMR Biomedicne, 23,* 803–820.

Jones, D. K., Knösche T. R, and Turner R. (2013). White matter integrity, fiber count, and other fallacies: The do's and don'ts of diffusion MRI. *NeuroImage, 73,* 239–254. doi:10.1016/j.neuroimage.2012.06.081

Jones, D. K., Travis, A. R., Eden, G., Pierpaoli, C., and Basser, P. J. (2005b). PASTA: Pointwise assessment of streamline tractography attributes. *Magnetic Resonance in Medicine, 53*(6), 1462–1467. doi:10.1002/mrm.20484

Kanaan, R. A., Barker, G. J., Catani, M., Howard, R., Jones, D. K., McGuire, P. K., Ng, V. W., and Shergill, S. S. (2006). Tract-specific anisotropy measurements in diffusion tensor imaging. *Psychiatry Research, 146*(1), 73–82.

Kim, S., Pickup, S., and Poptani, H. (2010). Effects of cardiac pulsation in diffusion tensor imaging of the rat brain. *Journal of Neuroscience Methods, 194*(1), 116–121. http://doi.org/10.1016/j.jneumeth.2010.10.003

Koch, M. and Norris, G. (2010). Artifacts and pitfalls in diffusion MR imaging. In J. H. Gillard, A. D. Waldman, and P. B. Barker (Eds.), *Clinical MR Neuroimaging.* New York, NY: Cambridge University Press.

Kreilkamp, B. A., Zacà, D., Papinutto, N., and Jovicich, J. (2016). Retrospective head motion correction approaches for diffusion tensor imaging: Effects of preprocessing choices on biases and reproducibility of scalar diffusion metrics. *Journal of Magnetic Resonance Imaging, 43*(1), 99–106. doi:10.1002/jmri.24965

Le Bihan, D., Poupon, C., Amadon, A., and Lethimonnier, F. (2006). Artifacts and pitfalls in diffusion MRI. *Journal Magnetic Resonance Imaging, 24,* 478–488. doi:10.1002/jmri.20683

Li, X., Yang, J., Gao, J., Luo, X., Zhou, Z., Hu, Y. et al. (2014). A robust post-processing workflow for datasets with motion artifacts in diffusion kurtosis imaging. *PLOS One, 9*(4), e94592. doi:10.1371/journal.pone.0094592

Mansfield, P. (1977). Multi-planar image formation using NMR spin echoes. *Journal of Physics C,* 10, 55–58.

Marchini, J. L., and Smith, S. M. (2003). On bias in the estimation of autocorrelations for fMRI voxel time-series analysis. *NeuroImage, 18*(1), 83–90. doi:10.1006/nimg.2002.1321

Mohammadi, S., Möller, H. E., Kugel, H., Müller, D. K., and Deppe, M. (2010). Correcting eddy current and motion effects by affine whole-brain registrations: Evaluation of three-dimensional distortions and comparison with slicewise correction. *Magnetic Resonance in Medicine, 64*(4), 1047–1056.

Morgan, P. S., Bowtell, R. W., Mcintyre, D. J., and Worthington, B. S. (2004). Correction of spatial distortion in EPI due to inhomogeneous static magnetic fields using the reversed gradient method. *Journal of Magnetic Resonance Imaging, 19*(4), 499–507. doi:10.1002/jmri.20032

Norris, D. G. (2001). Implications of bulk motion for diffusion weighted imaging experiments: Effects, mechanisms, and solutions. *Journal of Magnetic Resonance Imaging, 13*, 486–495.

Nunes, R. G., Jezzard, P., and Clare, S. (2005). Investigations on the efficiency of cardiac-gated methods for the acquisition of diffusion-weighted images. *Journal of Magnetic Resonance, 177*(1), 102–110. doi:10.1016/j.jmr.2005.07.005

Papadakis, N. G., Martin, K. M., Pickard, J. D., Hall, L. D., Carpenter, T. A., and Huang, C. L. (2000). Gradient preemphasis calibration in diffusion-weighted echo-planar imaging. *Magnetic Resonance in Medicine, 44*(4), 616–624. doi:10.1002/1522-2594(200010)44:4<616::aid-mrm16>3.0.co;2-t

Papadakis, N. G., Smponias, T., Berwick, J., and Mayhew, J. E. (2005). K-space correction of eddy-current-induced distortions in diffusion-weighted echo-planar imaging. *Magnetic Resonance in Medicine, 53*(5), 1103–1111. doi:10.1002/mrm.20429

Pasternak, O., Sochen, N., Gur, Y., Intrator, N., and Assaf, Y. (2009). Free water elimination and mapping from diffusion MRI. *Magnetic Resonance in Medicine, 62*(3), 717–730. doi:10.1002/mrm.22055

Peterson, D. J., Landman, B. A, and Cutting, L. E. (2008). The impact of robust tensor estimation on voxel-wise analysis of DTI data. *Proceedings of the International Society for Magnetic Resonance in Medicine Scientific Meeting and Exhibition, 16*, 1823.

Pierpaoli, C. (2011). Artifacts in diffusion MRI. In D. K. Jones (Ed.), *Diffusion MRI Theory, Methods, and Applications.* New York, NY: Oxford University Press.

Pierpaoli C., Walker L., Irfanoglu M. O., Barnett A., Basser P., Chang L.-C., Koay C., Pajevic S., Rohde G., Sarlls J., and Wu M. (2010). TORTOISE: an integrated software package for processing of diffusion MRI data. ISMRM 18th annual meeting, Stockholm, Sweden.

Pipe, J. G., Farthing, V. G., and Forbes, K. P. (2002). Multishot diffusion-weighted FSE using PROPELLER MRI. *Magnetic Resonance in Medicine, 47*(1), 42–52. doi:10.1002/mrm.10014

Pipe, J. G. and Zwart, N. (2006). Turboprop: Improved PROPELLER imaging. *Magnetic Resonance in Medicine, 55*(2), 380–385. doi:10.1002/mrm.20768

Reese, T., Heid, O., Weisskoff, R., and Wedeen, V. (2003). Reduction of eddy-current-induced distortion in diffusion MRI using a twice-refocused spin echo. *Magnetic Resonance in Medicine, 49*(1), 177–182. doi:10.1002/mrm.10308

Rohde, G. K., Barnett, A. S., Basser, P. J., Marenco, S., and Pierpaoli, C. (2004). Comprehensive approach for correction of motion and distortion in diffusion-weighted MRI. *Magnetic Resonance in Medicine, 51*(1), 103–114.

Schmithorst, V. J. and Dardzinski, B. J. (2002). Automatic gradient preemphasis adjustment: A 15-minute journey to improved diffusion-weighted echo-planar imaging. *Magnetic Resonance in Medicine, 47*(1), 208–212. doi:10.1002/mrm.10022

Schwarz, C. G., Reid, R. I., Gunter, J. L., Senjem, M. L., Przybelski, S. A., Zuk, S. M. et al. (2014). Improved DTI registration allows voxel-based analysis that outperforms tract-based spatial statistics. *NeuroImage, 94*, 65–78. doi:10.1016/j.neuroimage.2014.03.026

Shergill, S., Kanaan, R. A., Chitnis, X. A., O'Daly, O., Jones, D. K, Frangou, S. et al. (2007). A diffusion tensor imaging study of fasciculi in schizophrenia. *American Journal of Psychiatry, 164*(3), 467. doi:10.1176/appi.ajp.164.3.467

Smith, S. M., Behrens, T. E., Ciccarelli, O., Cader, M. Z., Jenkinson, M., Johansen-Berg, H. et al. (2006). Tract-based spatial statistics: Voxelwise analysis of multi-subject diffusion data. *NeuroImage, 31*(4), 1487–1505.

Studholme, C., Hill, D. L., and Hawkes, D. J. (1999). An overlap invariant entropy measure of 3D medical image alignment. *Pattern Recognition, 32,* 71–86.

Tofts, P. S., Lloyd, D., and Clark, C. A., Barker, G. J., Parker, G. J., McConville, P. et al. (2000). Test liquids for quantitative MRI measurements of self-diffusion coefficient in vivo. *Magnetic Resonance in Medicine, 43,* 368–374.

Triantafyllou, C., Hoge, R., Krueger, G., Wiggins, C., Potthast, A., Wiggins, G., and Wald, L. (2005). Comparison of physiological noise at 1.5 T, 3 T and 7 T and optimization of fMRI acquisition parameters. *NeuroImage, 26*(1), 243–250. doi:10.1016/j.neuroimage.2005.01.007

Tuch, D. S. (2004). Q-ball imaging. *Magnetic Resonance in Medicine,* 52 6, 1358–1372.

Tuch, D. S., Reese, T. G., Wiegell, M. R., Makris, N., Belliveau, J. W., and Wedeen, V. J. (2002). High angular resolution diffusion imaging reveals intravoxel white matter fiber heterogeneity. *Magnetic Resonance in Medicine, 48*(4), 577–582. doi:10.1002/mrm.10268

Turner, R., Bihan, D. L., Maier, J., Vavrek, R., Hedges, L. K., and Pekar, J. (1990). Echo-planar imaging of intravoxel incoherent motion. *Radiology, 177*(2), 407–414. doi:10.1148/radiology.177.2.2217777

Vavrek, R. M. and MacFall, J. R. (1995). Hardware considerations for diffusion/perfusion imaging. In D. Le Bihan (Ed.), Diffusion and Perfusion Magnetic Resonance Imaging. *Applications to Functional MRI* (pp. 67–72). New York, NY: Raven Press.

Walker, L., Chang, L., Koay, C. G., Sharma, N., Cohen, L., Verma, R., and Pierpaoli, C. (2011). Effects of physiological noise in population analysis of diffusion tensor MRI data. *NeuroImage, 54*(2), 1168–1177. doi:10.1016/j.neuroimage.2010.08.048

Wright, A. C., Bataille, H., Ong, H. H., Wehrli, S. L., Song, H. K., and F. W. Wehrli J. (2007). Construction and calibration of a 50 T/m z-gradient coil for quantitative diffusion microimaging. *Journal of Magnetic Resonance,* 186(1), 17–25.

Yoshiura, T., Wu, O., Zaheer, A., Reese, T. G., and Sorensen, A. G. (2001). Highly diffusion-sensitized MRI of brain: Dissociation of gray and white matter. *Magnetic Resonance in Medicine, 45*(5), 734–740. doi:10.1002/mrm.1100

Perfusion MR Imaging

3.1 Introduction

Focus Point

- Perfusion is the steady-state delivery of blood to an element of tissue.
- A bolus (from Latin *bolus*, i.e., ball) is the administration of a discrete amount of contrast agent in order to raise its concentration in blood to an effective level.
- Quantitative estimation of cerebral perfusion parameters using a bolus injection can be achieved either by
  - Emphasizing the gadolinium-based contrast agent's susceptibility effects through T2* or T2-weighted imaging, usually referred to as dynamic suscepti-bility contrast-enhanced MR imaging (DSC-MRI)
  - Emphasizing the gadolinium-based contrast agent's T1 relaxivity properties using T1-weighted dynamic contrast-enhanced MR imaging (DCE-MRI) to better characterize BBB leakage
  - Using magnetically labeled arterial blood water as a flow tracer rather than gadolinium-based contrast agent to estimate perfusion, usually referred to as arterial spin labeling (ASL)

The imaging of tissue microvasculature is of great importance and one of the key parameters for the evaluation and monitoring of brain pathology, providing an invaluable supplement to classical imaging as well as to the other advanced magnetic resonance (MR) techniques.

Perfusion MR imaging (MRI) is currently an evolving technology that has become a popular alternative to ionizing radiation techniques, such as nuclear medicine and computed tomog-raphy (CT)-based perfusion, for the study of cerebral hemodynamics and blood flow. These techniques have been inspirational regarding brain perfusion, but are now mainly limited by the absorbed radiation dose restrictions, especially concerning repeated exams, as well as the low spatial resolution compared to MRI.

"Perfusion" is defined as the steady-state delivery of blood to an element of tissue, such as the capillary blood flow (Welker et al., 2015). In MR perfusion imaging, the idea is either to use an exogenous, intravascular, nondiffusible contrast agent such as a gadolinium-based contrast agent (GBCA) and to analyze the quantitative cerebral perfusion parameters by exploiting the signal change through a series of dynamic images following its bolus administration, or to use the patient's own water molecules as an endogenous diffusible tracer.

Hence, cerebral perfusion MRI can be carried out using three techniques, two with contrast agent and one without. These are as follows:

1. The more popular and widely accepted T2 or T2* weighted dynamic susceptibility contrast (DSC) first-pass imaging technique to estimate blood volume and tumor neovasculature
2. The T1-weighted dynamic contrast enhanced (DCE) MRI technique to estimate capillary permeability
3. The arterial spin labeling (ASL) MRI technique for the detection of regional blood flow, using magnetically labeled arterial blood water as a contrast agent rather than a gadolinium-based agent

Furthermore, several investigators are currently exploring advanced techniques such as first-pass permeability estimation, and vascular morphologic imaging (Young and Qu, 2016), nevertheless these techniques, although extremely promising, have not yet been validated sufficiently for routine clinical use. Eventually, the noninvasive nature of MRI, accompanied with the very high resolution that can be achieved with the modern high magnetic field strength scanners allows the integration of these very promising tools into the advanced neuroimaging protocols.

3.2 DSC MRI

One of the most widely employed techniques for evaluating cerebral perfusion imaging is the DSC method, which utilizes the susceptibility effects of the contrast agent using T2- or T2*-weighted images. The idea is to monitor (or "track") the first pass of a concentrated intravascular bolus of paramagnetic Gd contrast agent by a dynamic series of MR images, exploiting the signal change due to susceptibility effects.

A bolus (from Latin *bolus,* i.e., ball) is the administration of a discrete amount of drug or other compound (this case a gadolinium-based contrast agent) in order to raise its concentration in the blood to an effective level.

> Please note, for the production of reliable perfusion results the rapid, high-pressure injection ("bolus") of a high concentration contrast dose (typical value: 0.2 mmol/kg) at a high rate (typical value ≥5 mL/s) must be used.

The analysis of the varying intensity of T2*-weighted images can be used to provide quantitative hemodynamic maps. This time course of the MRI signal as the gadolinium passes through a patient's brain is illustrated in Figure 3.1. A 60-year old man presented with an atypical meningioma is depicted on Figure 3.1a, an axial T1-weighted image post contrast and 3.1b, the parametric cerebral blood volume (CBV) map in the intratumoral (region of interest [ROI] 1) and peritumoral area (ROI 2). The derived signal intensity-time curves for each ROI (green curve) are shown in the small boxes.

The data acquisition in the DSC technique typically consists of rapid measurements of the MR signal change using a single-shot echo planar imaging (EPI) sequence, covering the whole brain every 1–2 sec, which is continuously repeated for 40–60 sec, thus producing a total of about 30 images. Therefore, it is usually also referred to as the "bolus-tracking technique." The prerequirement is to allow for a number of images before the arrival of the bolus in the brain and end the acquisition 10–20 sec after recirculation of the first pass of the contrast agent.

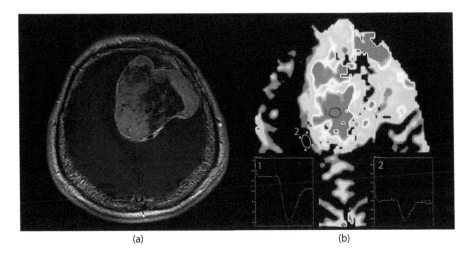

(a)   (b)

FIGURE 3.1   A 60-year-old man presented with an atypical meningioma is depicted on (a) an axial T1-weighted image post contrast and (b) the parametric cerebral blood volume (CBV) map in the intratumoral (ROI 1) and peritumoral area (ROI 2). The derived signal intensity-time curves for each ROI (green curve) are shown in the boxes.

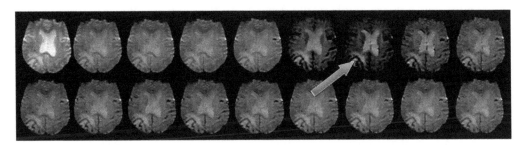

FIGURE 3.2   The dynamic effect of contrast flow (bolus) in a T2* EPI sequence. The signal intensity decreases significantly during the peak of the bolus (arrow).

Following the bolus injection of the contrast agent into the bloodstream, local susceptibility gradients will be generated, which will reduce the T2* relaxation time yielding a transient loss in signal intensity in the images as the contrast material passes through the vasculature and capillaries (Holdsworth and Bammer, 2008). The dynamic effect of contrast flow in this rapid gradient echo EPI sequence is illustrated in Figure 3.2. For illustration purposes only about half of the T2* images have been used.

These dynamic T2*-weighted signal changes (or "time-series") can be postprocessed into time-signal intensity curves for each voxel (as seen in Figure 3.1, small boxes), and then converted into the so-called concentration time curves, which will be analyzed in detail in the next section. With appropriate knowledge of the arterial input function (AIF), and by applying kinetic analysis, important quantitative physiologic parameters can be derived, such as parametric maps of cerebral blood flow (CBF), cerebral blood volume, and mean transit time (MTT), which are the most widely used. Moreover, based on the assumption that T2 or T2* relaxivity is directly proportional to contrast medium concentration, the principles of indicator dilution theory can be applied (Welker et al., 2015), and a number of other useful parameters can also be obtained as featured in Table 3.1.

TABLE 3.1   DSC-MRI Perfusion Parameters and Abbreviations Explained

Parameter	Abbreviation	Description
Cerebral blood volume	CBV	The volume of blood in a given region of brain tissue (measured in milliliters per 100 g of brain tissue).
Cerebral blood flow	CBF	The volume of blood passing through a given region of brain tissue per unit of time (measured in milliliters per minute per 100 g of brain tissue).
Mean transit time	MTT	The average time (measured in seconds) for blood to pass through a given region of brain tissue.
Time to peak	TTP	The time at which the minimal signal intensity (greatest signal loss) is reached (i.e., the point at which the GBCA concentration is maximal).
Time of maximum	Tmax	The time at which the CBF reaches its maximal value.
Peak height	Peak height	The maximal drop in signal intensity from precontrast baseline during the first-pass bolus phase of GBCA.
Percentage of signal recovery	PSR	The percentage of MR signal-intensity recovery relative to the precontrast baseline at the end of the first pass.

3.2.1 DSC Imaging Explained

Susceptibility contrast imaging basically depends on the paramagnetic properties of gadolinium, which substantially alters the average tissue bulk susceptibility. More analytically, in the presence of magnetic inhomogeneities due to the paramagnetic substance, the transverse magnetization dephases, causing the loss of signal in tissue close to the vascular bed.

In other words, following contrast injection, the gadolinium enhanced vessels will behave like a group of magnetized cylinders altering the local magnetic field, influencing the adjacent water protons. It follows that this strong signal all the water protons would produce being on the same radiofrequency influenced by the unaltered magnetic field will now decrease since the water protons will now be influenced unevenly, radiating at different frequencies.

Using T2- or T2*-weighted imaging the passage of a bolus of gadolinium-based contrast can be followed, revealing aspects of the cerebral microvasculature, this way aiding a wide variety of clinical applications including evaluation and classification of tumors, evaluation of stroke regions, and characterization of several cerebrovascular diseases (Kennan and Jager, 2003).

3.2.2 DSC Perfusion Parameters: CBV, CBF, MTT

3.2.2.1 CBV

CBV is defined as the volume of blood in a given amount of brain tissue. To calculate the CBV, measurements of the MR derived signal-intensity-time curves (Figure 3.3a) must be converted to concentration-time curves (Figure 3.3b).

As shown in Figure 3.3, for every pixel in the brain, a curve that depicts the changes in signal intensity over time during the first pass and recirculation of the contrast bolus is created. Assuming a linear relationship between the paramagnetic agent concentration and the change in transverse relaxation rate (called the "assumption of linearity"), the changes in concentration can be calculated using the following equation:

$$C(t) = -k * \log\big(S(t)/S_0\big)/\mathrm{TE} \tag{3.1}$$

where S_0 is the precontrast spin-echo signal intensity of the baseline images averaged after the signal has reached its steady state (usually after one to four images), $S(t)$ is the spin-echo signal

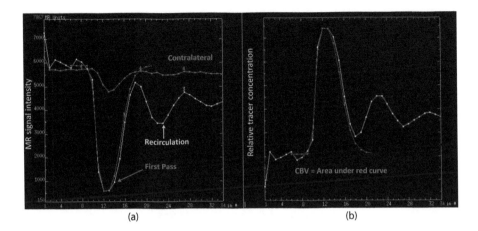

(a) (b)

FIGURE 3.3 (a) The MR signal-intensity-time curves of a lesion and the contralateral region. The recirculation curve is also included. (b) The relative concentration-time curve for CBV calculation.

intensity of the contrast at time t, and TE is the echo time. k is a scaling factor, which relates T2* changes to concentration in large vessels and parenchyma. This is because T2* changes can substantially differ between large vessels and parenchyma, and depend upon the overall concentration level and the vessel orientation to the main magnetic field (Rosen et al., 1990; Rosen et al., 1991). A similar equation applies for gradient echo acquisitions.

The assumption of linearity is widely used in perfusion measurements and has been confirmed by simulations and measurements in vivo (Simonsen et al., 1999). Nevertheless, it has also been shown that the physics of MR signal formation in perfused tissues can be complicated (Kiselev and Posse, 1998, 1999) and the linearity assumption may not apply for all contrast agent concentrations or tissues, causing overestimation of perfusion estimates (Kiselev, 2001). Hence, the commonly assumed linear relationship between contrast agent concentration and the T2* relaxation rate in arterial blood is now known to be invalid. A more recent study by Calamante et al., (2009), characterized the associated perfusion limitations, concluding that large errors were found when the linear assumption is used and that these can be greatly reduced when using a quadratic model, and quantifying perfusion as a relative measure. Hence, the postprocessing of the relationship between the T2*-weighted signal and tracer concentration may yield an estimate of the CBV on a pixel-by-pixel basis, by integrating the area under the contrast concentration curve (Figure 3.3b). It must be stressed at this point that the assumption here is that the contrast agent is only distributed in the plasma space (i.e., no capillary permeability) and that there is no recirculation of contrast. Moreover, tissue concentration must be normalized to the blood volume in a large vessel (i.e., 100% blood volume) such as the sagittal sinus.

CBV measurements can be performed using a ROI placement. The placement of a ROI within a lesion and a contralateral equally sized ROI within normal-appearing brain parenchyma can be used to calculate the so-called relative CBV (rCBV) and the normalized CBV (nCBV) (i.e., the ratio between the two ROIs) (Cha, 2006; Cha et al., 2002). The ROI placement on the lesion and contralateral area with the associated time curves is depicted in Figure 3.4. The resolution limit of perfusion weighted imaging using commercial software remains at 2–4 mm in plane voxel dimensions, despite the ongoing work by many investigators (Lupo et al., 2006). Therefore, the rCBV map can and should be overlaid on higher resolution anatomic images (T2 or T1 postcontrast) as shown in Figure 3.4, so optimum interpretation can be accomplished.

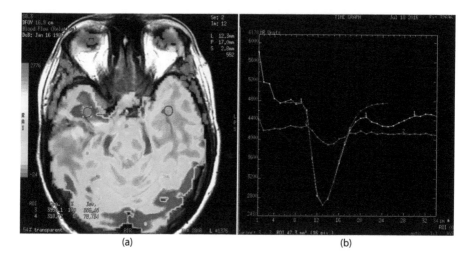

(a)   (b)

FIGURE 3.4   This figure illustrates (a) the region of interest (ROI) placement on the lesion and contralateral area and (b) their associated relaxivity time curves. The rCBV map in (a) is overlaid on a higher resolution anatomic image to facilitate optimum interpretation. The map is at ~50% transparency relative to the anatomical image.

CBV versus *rCBV*

CBV is a qualitative metric derived by integrating the relaxivity time curves as previously mentioned. Please note that this is not the absolute CBV, but rather a relative measure. Absolute volume can only be calculated by including an adjustment of the AIF (Covarrubias et al., 2004), with two pharmacokinetic conditions: (1) no recirculation of contrast and (2) no capillary permeability.

The recirculation of contrast can be corrected for by assessing only the first pass of gadolinium, while the second condition is more problematic as permeability can vary substantially. Due to this difficulty, the rCBV (i.e., relative to an internal control) is the most common metric reported in the literature, especially regarding tumors (Bobek-Billewicz et al., 2010; Sugahara et al., 2000). rCBV may be calculated relative to the contralateral normal white matter or an AIF, and it is reported without units as it is just a ratio.

There is also the quantitative CBV (qCBV); this can be calculated using specific techniques, with age-matched normative data and available postprocessing methods (Essig et al., 2013), but may significantly vary between white and gray matter. This is why rCBV is generally suggested as a more meaningful technique compared to absolute quantification (Bedekar et al., 2010; Calamante, 2013; Essig et al., 2013).

3.2.2.2 CBF

Likewise, the volume of blood passing through a given amount of brain tissue per unit time, defined as CBF, can be calculated. It is measured in milliliters per minute per 100 g of brain tissue and is a highly useful parameter to evaluate tissue fate. Nevertheless, in most instances, CBF is difficult to measure in absolute terms, and relative CBF (rCBF, relative to normal white matter) is typically used. As with CBV, a quantitative evaluation of CBF (qCBF) can only be performed using specific techniques, and it varies significantly between white and gray matter.

3.2.2.3 MTT

The third parameter that nicely describes the hemodynamic status of brain tissue is MTT. MTT is the average time (measured in seconds) for an infinitesimally small amount of the contrast agent to pass through a given region of the capillary system, or the time spent within a determinate volume of capillary circulation. Contrary to CBV or CBF, the advantage of MTT is that MTT maps in healthy subjects appear without substantial gray/white matter differences, therefore any abnormality will be easily noticeable. The central volume theorem (Dawson and Blomley, 1996) directly relates all three aforementioned parameters (CBV, CBF, and MTT) using the physical principles of the indicator dilution theory, so MTT can be calculated as follows:

$$MTT = CBV/CBF \qquad (3.2)$$

Unfortunately, the bolus duration is quite short (normally 1–2 seconds), and therefore a deconvolution approach is typically applied to correct for the effect of the bolus width, from which CBF can also be calculated (Holdsworth and Bammer, 2008).

3.3 DCE-MRI

The second contrast-based method is the DCE-MRI. Similar to DSC-MRI, a bolus of contrast agent is injected and hemodynamic differences are detected. Contrary to DSC, DCE depends on the T1 relaxation time since hemodynamic signals increase because of the T1 shortening effect associated with the concentrated intravascular bolus of paramagnetic Gd contrast agent. Therefore, it is usually also referred to as "T1 perfusion MRI" or "permeability MRI."

DCE-MRI involves the rapid and repeated T1-weighted images to capture the dynamic induced T1 signal changes in the tissue before, during, and after the intravenous administration of the contrast agent. Since the contrast media diffuses from the blood into the extravascular–extracellular space (EES) of tissue, T1-weighting is not affected by extravasation. It is the tissue perfusion as well as the permeability of the capillaries and their surface area that determine the diffusion flow rate. Hence, the presence of contrast medium causes shortening of the T1 relaxation rate and therefore a relaxivity-based method of tissue enhancement.

The quantification of contrast agent microvascular permeability utilizing a two-compartment pharmacokinetic model (plasma space and EES) yields important quantitative physiologic parameters such as the volume transfer constant related to the permeability–surface area (K^{trans}) and the fractional volume of the total plasma and EES (V_p and V_e, respectively). Another metric usually evaluated is the area under contrast curve (AUC).

The DCE quantification parameters with their description are featured in Table 3.2.

TABLE 3.2 DCE-MRI Perfusion Parameters and Abbreviations Explained

Parameter	Abbreviation	Description
Volume transfer constant	K^{trans}	Measure of microvascular permeability
Total plasma space volume	V_p	Fractional volume of the total plasma, and extravascular–
Total extravascular–extracellular space volume	V_e	extracellular space, based on pharmacokinetic model to quantify microvascular permeability
Rate constant	k_{ep}	$k_{ep} = K^{trans}/V_e$

3.3.1 DCE Imaging Explained

When a bolus of contrast agent is injected into the blood, its behavior in the tissue will be determined by two main factors besides perfusion: (1) the transfer of contrast across vessel walls, and (2) the diffusion of contrast in the interstitial space. In that sense, the DCE-MRI signal will be directly related to changes associated with the tissue microvasculature. The enhancement time line is usually characterized by an initial increase in the plasma concentration, which might be accompanied by some interstitial tissue leakage, and then plasma concentration will decrease due to physiological diffusion while tissue concentration will also show a delayed decrease depending on the underlying pathophysiology. The overall contrast enhancement behavior as a function of time is depicted in Figure 3.5.

This dynamic contrast enhancement curve can be analyzed in several components characterizing the different anatomic and physiological processes taking place. The total contrast enhancement curve consists of an initial increase correlated to total tissue blood flow while the curve's height can be correlated to the total blood volume. The rest of the curve consists of the interstitial tissue contrast leakage, indicative of capillary area and permeability function, followed by the tissue extracellular space and plasma interstitial volume.

A typical DCE curve in the brain with the corresponding K^{trans}, V_e, and initial AUC (iAUC) maps is depicted in Figure 3.6.

DCE signal enhancement data can be evaluated using three methods: (1) simple visual assessment of the concentration time curves, (2) signal intensity analysis by semi-quantitative descriptive indices, and (3) pharmacokinetic model-based quantification techniques.

Semi-quantitative parameters can be calculated without considering the physiological basis and include the AUC, iAUC (uptake part only), arrival time (AT), time to peak (TTP), wash-in rate (WIR), wash-out rate (WOR), maximum enhancement (ME), and percent maximum enhancement (%ME) compared to baseline (Tofts et al., 1999).

Nevertheless, it has to be stressed that these absolute values can only be used relative to signal intensity changes in other lesions or relative to the contralateral normal area (Jackson

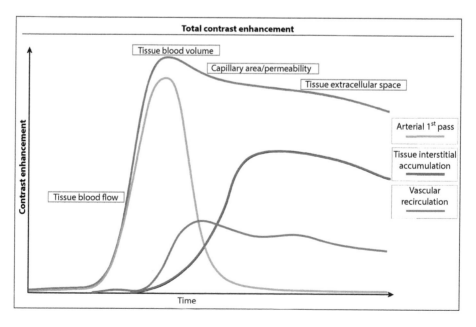

FIGURE 3.5   DCE contrast enhancement behavior as a function of time.

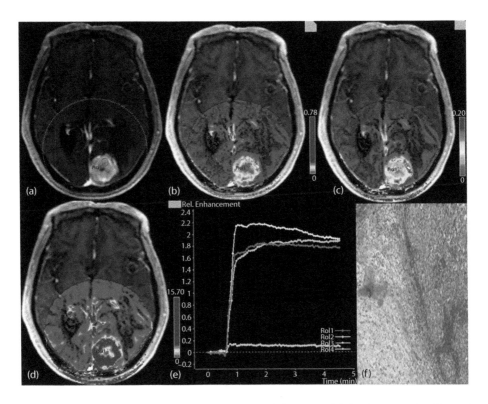

FIGURE 3.6 A metastatic tumor in a patient in the left occipital lobe from renal cell carcinoma. Contrast-enhanced T1-weighted MRI showed that the tumor is obviously enhanced with severe peritumoral edema (a) and the corresponding K^{trans} (b), Ve (c) and iAUC (d) maps showed high value inside tumor parenchyma except central necrosis; concentration-time curve of the tumor (e) manifested as a plateau; brain metastases from renal cell carcinoma (f), (Hematoxylin–Eosin (HE) × 4). (From Zhao, J. et al., *PLOS One*, 10, e0138573, 2015. With permission.)

et al., 2014). On the other hand, quantitative hemodynamic indices should be derived taking into account the physiological basis, and requires three steps: (1) conversion of signal intensity to gadolinium concentration; (2) selection of an appropriate tissue model; and (3) estimation of model parameters from the fitted data (Elster, n.d.).

Following the conversion of signal intensity data into Gd concentrations, the second step is to select an appropriate model to which these data will be fitted. Three major models (from Tofts, Larsson, and Brix) for collecting and analyzing DCE-MRI data were initially introduced (Tofts et al., 1999; Brix et al., 1991; Larsson et al., 1990). All aforementioned models use compartmental analysis to obtain the physiological parameters previously mentioned, representing the blood plasma and the abnormal EES. The parameters measured are: the permeability surface area product per unit volume of tissue, between plasma and EES; the EES fractional volume (V_e) and fractional plasma volume (V_p); and K_{ep}, the rate constant or reflux rate (K_{ep} = K^{trans}/V_e). One of the simplest and most widely used models is that proposed by Tofts et al. illustrated in Figure 3.7.

The most important and most frequently used DCE parameter is K^{trans}, the volume transfer constant for gadolinium between blood plasma and EES. K^{trans} depends on the delicate balance between capillary permeability and blood flow in the tissue of interest, and therefore reflects the sum of all underlying processes that determine the rate of gadolinium influx from plasma

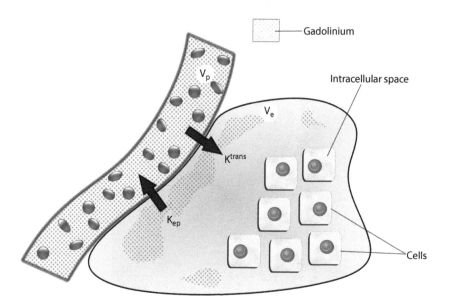

FIGURE 3.7 Tofts Model. The relative sizes of the extracellular spaces are exaggerated compared to that of the non-gadolinium-containing intracellular spaces. Moreover, the fractional plasma volume (V_p) is generally much smaller than the fractional tissue EES (V_e). (Adapted from and courtesy of Allen D. Elster, MRIquestions.com.)

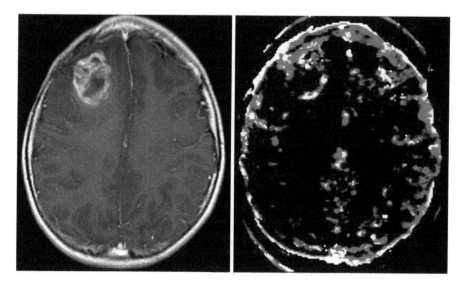

FIGURE 3.8 The postcontrast T1-weighted MR image of a malignant glioma (left) with the corresponding DCE K^{trans} map (right). (Courtesy Allen D. Elster, MRIquestions.com.)

into the EES. It is given in values of (1/time), or min^{-1}. K^{trans} correlates with the initial slope ("wash-in" rate) of the time-intensity curve. The K^{trans} map with the corresponding postcontrast T1-weighted MR image of a malignant glioma is depicted in Figure 3.8.

As already mentioned, there is also K_{ep}, which is the time constant for gadolinium reflux from the EES back into the vascular system. Both K^{trans} and K_{ep} have the same units (min^{-1}). The fractional volume of the EES, denoted as V_e, is dimensionless since it is defined as the volume of the EES per unit volume of tissue. A nice explanation given by Prof. D. Elster is that V_e can be considered to reflect the amount of "room" available within the tissue interstitium for accumulating the contrast agent. Taking V_e into account, it follows that the rate constant K_{ep} should always be greater than the transfer constant K^{trans} since $K_{ep} = K^{trans}/V_e$ and $V_e < 1$.

All the above mentioned parameters arise from the original Tofts model (Tofts and Kermode, 1991) depicted in Equation 3.1:

$$C_t(t) = K^{trans} \int_0^t C_p(\tau)^{-(K^{trans}/v_e)(t-\tau)} \, d\tau \tag{3.3}$$

Nevertheless, a common modification to a more general equation to express the hemodynamic event after injecting the contrast agent usually referred to as the extended Tofts model (Tofts, 1997) is

$$C_t(t) = v_p C_p(t) + K^{trans} \int_0^t C_p(\tau) e^{-(K^{trans}/v_e)(t-\tau)} \, d\tau \tag{3.4}$$

In this extended model, the difference is that flow and permeability limitations can be taken into account and thus K^{trans} can be interpreted even when the contrast flow in the tissue might be insufficient. At this case blood perfusion will determine contrast agent kinetics and K^{trans} cannot be characterized as a permeability parameter. On the other hand, if contrast flow in the tissue is sufficient without flow limitations, then K^{trans} is characterized as a permeability surface area product since the transferring of contrast across the vessel walls will determine contrast agent kinetics.

The last parameter in the expanded Tofts model is the fractional plasma volume or V_p. Its contribution is generally greater in highly vascular tumors where it can reach 10% of the total signal and should be taken into account. Nevertheless, it is evident that the use of a single parameter to evaluate both perfusion and capillary permeability might be problematic and therefore several investigators sought to develop models that would enable separate measurements (Ewing and Bagher-Ebadian, 2013; Hsu et al., 2013; Larsson et al., 2009; Sourbron et al., 2009).

However, a major limitation is that these models require relatively high temporal resolution (about 1 sec, while a typical temporal resolution currently used is 5 sec), therefore their use is still relatively confined.

Another limitation is that multiple models may be appropriate for a given tissue type and this is a typical problem associated with DCE parameter evaluation. Obviously, the results would then depend on the chosen model (Sourbron, 2010; Haroon et al., 2004), therefore an appropriate selection is a crucial parameter.

The easiest approach to tackle the model selection problem or a "general rule of thumb" is that the most appropriate model is the simplest one that provides a good fit to the data (Sourbron, 2010). In recent years, DCE-MRI is increasingly being used for a more accurate mapping of the blood volumes in a tumor lesion, hence it can be used for tumor grading, differential diagnosis procedures, evaluation of radiotherapy treatment effect, or for chemotherapy treatment response monitoring. Typical findings indicating increased tumor vascularity are increased AUC, peak enhancement, and K^{trans} values. In a recent publication Zhao, et al. (2015) quantitatively evaluated the diagnostic efficiency of DCE-MRI parameters combined with diffusion for the classification of brain tumors. They concluded that permeability parameters from Diffusion, and especially DCE-MR, may aid in the grading of gliomas as well as brain tumor discrimination, and that their use is strongly recommended.

3.4 ASL

The fundamental goal of blood perfusion in the human body is to provide tissue with oxygen and nutrients. Utilizing arterial water as an endogenous diffusible tracer (i.e., originating within the organism), it became possible to obtain absolute values of blood perfusion in the tissue without the use of an exogenous contrast agent. Unlike gadolinium contrast agents, arterial water is diffusible, so water molecules do not remain confined to the extravascular space but freely move from capillaries into tissue parenchyma, including cells. The technique was named ASL since it uses magnetic labeling of inflowing arterial blood to produce images. The main idea back in 1992 was to acquire two different images, one with labeling of the incoming arterial blood and one without labeling, and then subtract them (Detre et al., 1992; Detre et al., 1994; Williams et al., 1992). This signal difference was used to produce the ASL image. Obviously, the signal-to-noise ratio of ASL is relatively low, but on the other hand, this is a completely noninvasive technique using an endogenous rather than an exogenous contrast agent, with no radiation burden and, therefore, can be repeated as many times as necessary (Wong, 2014).

Especially comparing ASL to nuclear medicine techniques (positron emission tomography [PET] and single photon emission computed tomography [SPECT]) its main advantage is the improved spatial resolution, which is typically 4 mm even at 1.5 T. However, ASL's current main disadvantage is that the analysis methods remain quite complex and are still susceptible to motion and other artifacts. For more information about limitations please refer to the next chapter.

3.4.1 ASL Imaging Explained

The last decade has been a very exciting period for ASL techniques in MRI. A large number of studies have proved the potential usefulness of this wonderful and completely noninvasive technique in a number of physiological processes and diseases.

All current methods and recent developments are variations of the original proposed method by Detre and Williams in the early 1990s. Their method was based on the inversion of spins of the inflowing protons in the carotid arteries (called magnetic labeling), so that when entering the brain tissue in a labeled state, they would produce a magnetization change revealing aspects of the local microperfusion. In other words, any regional blood flow change from labeled protons would be associated with a difference in the image contrast. To accomplish that, a subtraction of a "labeled" image (or "label" image) from an "unlabeled" image (or "reference" image) would be required. The important aspect is that the exchange of the "labeled" protons with the brain tissue protons (which slightly alters the local magnetization), depends on the T1 relaxation time of the tissue of interest. That is, labeled blood, flows into capillary sites with the T1 of blood, diffusing in the tissue water space, and is replaced by blood from the capillary vessel with the T1 of tissue. Hence, the subtraction between the label and the reference image should theoretically produce a signal difference that is absolutely proportional to the diffusing "fresh" blood into the tissue of interest since all other effects should be canceled out.

Schematically the basic principle of ASL is illustrated in Figure 3.9.

3.4.2 Different ASL Techniques

Depending on the labeling method, ASL may be classified into four broad types: CASL (continuous ASL), pCASL (pseudo-continuous ASL), PASL (pulsed ASL), and VSASL (velocity-selective ASL).

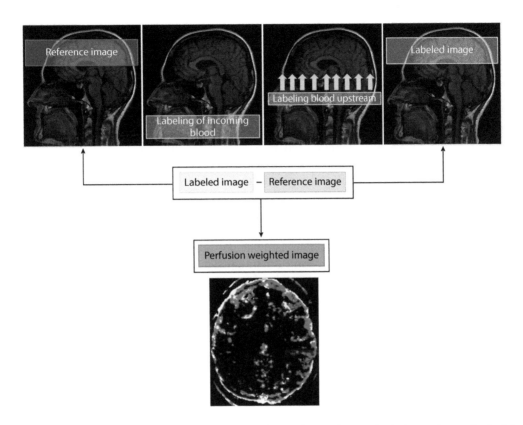

FIGURE 3.9 The basic principle of ASL. From left to right: a reference brain image is acquired without "labeling" and then "labeling" pulses are applied to invert the magnetization of water protons of inflowing blood. "Labeled" protons flow into the reference imaging volume and their exchange with the brain tissue protons slightly alters the local magnetization (approx. by 1%–2%). The reference volume is re-acquired (now called: labeled image). Finally, the labeled image is subtracted from the reference image producing a perfusion weighted image.

3.4.2.1 CASL and pCASL

In the earliest approach of ASL, which is CASL, the labeling of flowing proton spins was accomplished by continuous radiofrequency (RF) pulses along with magnetic field gradients in the direction of blood flow. It is usually referred to as "adiabatic inversion" or adiabatic fast passage (AFP) (Williams et al., 1992). (*An adiabatic process is one in which no heat is gained or lost by the system*). Although CASL could provide higher perfusion contrast compared to other labeling types (Wang et al., 2002), this approach had obvious limitations, especially related to a high energy deposition due to the continuous RF application required to bring about the adiabatic inversion (high SAR), as well as hardware and experimental restrictions related to the long labeling pulse (Borogovac and Asllani, 2012). Therefore, more recently, it has been replaced by the so-called pCASL, which is a hybrid approach that uses a long series of short RF-pulses using the body coil, together with a strong slice-selection gradient. pCASL thus combines favorable features of CASL (high signal-to-noise) with those of PASL (lower energy deposition) (Silva and Kim, 1999; Elster, n.d.).

Other approaches to overcome the problem of long RF pulses include the dual coil CASL (dc-CASL), which uses two separate coils for labeling and imaging, respectively, and almost continuous ASL (ACASL), where the labeling pulse is regularly and briefly interrupted in order to reduce SAR (Zaharchuk et al., 1999, 2007).

3.4.2.2 PASL

In contrast to CASL techniques, PASL involves more easily implemented short RF pulses (10–15 ms), which result in the inversion of the blood in a specific region adjacent to the imaging volume usually referred to as the inversion slab (Golay et al., 2004). PASL techniques are further divided based on the labeling method used, into those in which the label spins symmetrically with respect to the plane of imaging like flow-sensitive alternating inversion recovery (FAIR) (Kim, 1995), or asymmetrically like echo-planar imaging-based signal targeting by alternating radiofrequency pulses (EPISTAR) (Edelman et al., 1994) and proximal inversion with control of off-resonance effects (PICORE) (Wong et al., 1997).

FAIR involves two inversion recovery acquisitions, one used for the magnetization of the selected slice (i.e., slice-selective inversion) and one not. After the delay, the tissue magnetization of the nonselective inversion is assumed to be equal to that of the tissue. In EPISTAR and PICORE, imaging slice saturation and proximal inversion pulses are used with fast imaging (EPI) after a short delay following the inversion of magnetization. An additional control image is similarly acquired in a symmetrical slab of tissue distal to the imaging slice.

The choice between FAIR, or EPISTAR and PICORE methods mainly depends on the expected inflow geometry of blood to the imaging slab. EPISTAR and PICORE is the reasonable choice for axial brain perfusion imaging when inflow of blood is coming for the carotid arteries as the labeling takes place from one side only. FAIR tags both sides of the imaging volume and may be a better choice for sagittal brain perfusion imaging where inflow may be from multiple directions. On the other hand, although FAIR is easy to implement and relatively straightforward in its application, it may be more sensitive to contamination of venous inflow, and multi-slice mode remains problematic due to inversion slice-profile imperfections (Kim et al., 1997).

3.4.2.3 VSASL

Finally, there is the VSASL. As is suggested by its name, while VSASL is based on all the same elements of conventional ASL label and control images, it uses a very different approach as labeling is based on the velocity of arterial water rather than on its spatial location (as in CASL and PASL). Using binominal velocity selective pulses, the labeling is restricted to those spins whose velocity meets a certain cut-off value. Assuming a monotonic decrease in the velocities of the arterial tree, the amount of labeled blood in a given imaged voxel will be analogous to the amount of blood flow and delay time of that voxel. Hence, theoretically, certain spins whose velocity meets the criteria can be tagged. This tagging will not be spatially selective, and therefore it can be made sensitive to different blood velocities avoiding transit time problems. Generally, a relatively low cut-off value of 4 cm/s is recommended (Wu et al., 2007).

Inevitably, the question arises: Which ASL technique is better?

The question is obviously difficult to answer, but the choice should depend in part on the associated application and brain coverage issues, as well as the availability of hardware, software, and technical expertise. Generally, there has been a preference of PASL over CASL in the literature since it is easier to implement and conceptually more straightforward.

Moreover, PASL is superior to CASL considering the SAR and magnetization transfer issues. On the other hand, CASL yields higher SNR and less sensitivity to transit times. Hence, pCASL, which is the hybrid method between CASL and PASL, has become the most popular method and has contributed to a substantial increase in broad range ASL applications at high field MRI (3T). This is mainly due to its higher efficiency and capability for multi-slice imaging. Last, although VSASL sounds promising, further investigation is needed to avoid biased comparisons (Duhamel et al., 2003).

3.4.3 ASL beyond CBF Estimation

Beyond CBF evaluation, ASL may be used to investigate a number of physiological parameters related to the vascular supply of tissue such as angiography (Robson et al., 2010) quantification of the blood volume of the arterioles and capillaries prior to the diffusion of the tracer out of the microvasculature (Petersen et al., 2006), and characterization of the blood-brain-barrier breakdown with vascular permeability estimation (Li et al., 2005).

3.5 Conclusions and Future Perspectives

Tissue perfusion is a fundamental biological function for the delivery of oxygen and nutrients in the tissues, and its analytical study in the brain provides functional information about the underlying pathophysiology in a great number of disorders, including the evaluation and classification of tumors, stroke evaluation, and characterization of other diseases and their prognoses. More specifically, perfusion techniques can provide information regarding tumor cellularity, tissue invasion, metabolism, and microvasculature, and therefore their contribution towards a multi-parametric brain tumor evaluation is very important.

Perfusion MRI techniques can be classified according to the contrast mechanism used, into exogenous and endogenous. DSC and DCE are exogenous methods that include the use of a bolus contrast agent. They both generally provide high sensitivity and spatial resolution, and are therefore widely used in clinical applications. More specifically, DSC is faster than DCE and can be easily used for whole brain coverage, while DCE is better for the estimation of blood volume and produces measures of capillary permeability.

On the other hand, the ASL method provides the unique opportunity to evaluate hemo-dynamic information without requiring an exogenous contrast agent, using tagged arterial water protons. Over the past several years, many technical advances in scanner hardware and software (multichannel coils, parallel imaging, etc.) have enabled the increase of ASL's clinical and research applications, including applications in children as a completely noninvasive and repeatable technique with better accuracy for quantification.

Although a number of limitations (e.g., standardization of techniques and improvement of postprocessing software) in the different endogenous or exogenous contrast perfusion methods remain, the acquired parameters are of great importance and the inclusion of perfusion MRI in the clinical routine is expected to improve diagnostic accuracy and understanding of tumor pathophysiology.

Table 3.3 is an attempt to summarize and comprehensively describe perfusion MRI in its current state. It includes all three types of perfusion MRI techniques, with the associated sources of signals, imaging sequences and common imaging parameters, as well as the quantifications and major theoretical assumptions, and can be used as a quick reference.

TABLE 3.3   Current Status of Perfusion MRI

	DSC (Dynamic susceptibility contrast)	DCE (Dynamic contrast enhanced)	ASL (Arterial spin labeling)
Bolus material	Intravenous bolus injection of Gd-based contrast agent	Intravenous bolus injection of Gd-based contrast agent	Without contrast agent Use of arterial water
Contrast method	Bolus tracking	Bolus passage	Bolus labeling
Injection rate	>5 mL/sec	~3 mL/sec	-
Acquisition	First pass of contrast agent	Accumulation of contrast agent	Accumulation of labeled blood
Exogenous or endogenous method	Exogenous	Exogenous	Endogenous
Basic phenomenon evaluated	T2/T2* relaxation change	T1 relaxation change	Magnetic labeled blood T1 relaxation
Basic mechanism	• Use of paramagnetic material • Distortion of local magnetic field due to susceptibility effects • Reduction of T2/T2* around vessel • Result: Signal drop (dephasing and signal loss due to diffusion)	• Use of paramagnetic material induces "dipole-dipole" interactions • Reduction of T1 • Result: Signal enhancement (T1 shortening)	• Labeling of arterial blood (usually inversion) • Exchange of tagged blood in brain tissues from capillary vessel • Result: Signal drop (due to magnetization exchange)
Imaging types / sequences	T2 or T2*-weighted rapid GRE or EPI imaging	T1-weighted rapid GRE or EPI imaging	Short TE and long TR sequences (EPI/FSE)
Acquisition	2D or 3D dynamic acquisition	2D or 3D dynamic acquisition	Custom interleaved 2D or 3D EPI
Acquisition time	Short (1–2 min)	Long (>5 min)	Intermediate (3–5 min)
Temporal resolution	1–1.2 sec	5–10 sec	4–8 sec
Assumptions and considerations	• Intact BBB • Stable flow during the measurement • Negligible T1 effects • No recirculation of tracer	• Instantaneous passage of contrast agent	• No inflow and outflow effects
Quantitative parameters	• Absolute CBF (?) • Relative CBF • CBV, MTT	• K^{trans} • k_{ep} • v_e • v_p	• Absolute CBF

Note: BBB = brain-blood-barrier; CBF = cerebral blood flow; CBV = cerebral blood volume; EPI = echo planar imaging; FSE = fast spin echo; Gd = gadolinium; GRE = gradient echo; K^{trans} = volume transfer constant; MTT = mean transit time; k_{ep} = rate constant; TE = echo time; TR = repetition time; TTP = time-to-peak; V_e = total extravascular–extracellular space volume; V_p = total plasma space volume.

References

Bedekar, D., Jensen, T., and Schmainda, K. M. (2010). Standardization of relative cerebral blood volume (rCBV) image maps for ease of both inter- and intrapatient comparisons. *Magnetic Resonance in Medicine*, 64(3), 907–913. doi:10.1002/mrm.22445

Bobek-Billewicz, B., Majchrzak, H., Stasik-Pres, G., and Zarudzki, L. (2010). Differentiation between brain tumor recurrence and radiation injury using perfusion, diffusion-weighted imaging and MR spectroscopy. *Folia Neuropathologica*, 48(2), 81–92.

Borogovac, A. and Asllani, I. (2012). Erratum to Arterial spin labeling (ASL) fMRI: Advantages, theoretical constrains and experimental challenges in neurosciences. *International Journal of Biomedical Imaging*, 2012, 1–1. doi:10.1155/2012/658101

Brix, G., Semmler, W., Port, R., Schad, L. R., Layer, G., and Lorenz, W. J. (1991). Pharmacokinetic parameters in CNS Gd-DTPA enhanced MR imaging. *Journal of Computer Assisted Tomography*, 15(4), 621–628. doi:10.1097/00004728-199107000-00018

Calamante, F. (2013). Arterial input function in perfusion MRI: A comprehensive review. *Progress in Nuclear Magnetic Resonance Spectroscopy*, 74, 1–32. doi:10.1016/j.pnmrs.2013.04.002

Calamante, F., Connelly, A., and Osch, M. J. (2009). Nonlinear ΔR*2 effects in perfusion quantification using bolus-tracking MRI. *Magnetic Resonance in Medicine*, 61(2), 486–492. doi:10.1002/mrm.21839

Cha, S. (2006). Dynamic susceptibility-weighted contrast-enhanced perfusion MR imaging in pediatric patients. *Neuroimaging Clinics of North America*, 16(1), 137–147. doi:10.1016/j.nic.2005.11.006

Cha, S., Knopp, E. A., Johnson, G., Wetzel, S. G., Litt, A. W., and Zagzag, D. (2002). Intracranial mass lesions: Dynamic contrast-enhanced susceptibility-weighted echo-planar perfusion MR imaging. *Radiology*, 223(1), 11–29. doi:10.1148/radiol.2231010594

Covarrubias, D. J., Lev, M. H., and Rosen, B. R. (2004). Dynamic magnetic resonance perfusion imaging of brain tumors. *The Oncologist*, 9(5), 528–537.

Dawson, P. and Blomley, M. J. (1996). Contrast media as extracellular fluid space markers: Adaptation of the central volume theorem. *The British Journal of Radiology*, 69(824), 717–22. doi:10.1259/0007-1285-69-824-717

Detre, J. A., Grandis, D. J., Koretsky, A. P., Leigh, J. S., Roberts, D. A., Silva, A. C. et al. (1994). Tissue specific perfusion imaging using arterial spin labeling. *NMR in Biomedicine*, 7(1–2), 75–82.

Detre, J. A., Koretsky, A. P., Leigh, J. S., and Williams, D. S. (1992). Perfusion imaging. *Magnetic Resonance in Medicine*, 23(1), 37–45.

Duhamel, G., Bazelaire, C. D., and Alsop, D. C. (2003). Evaluation of systematic quantification errors in velocity-selective arterial spin labeling of the brain. *Magnetic Resonance in Medicine*, 50(1), 145–153. doi:10.1002/mrm.10510

Edelman, R. R., Siewert, B., Darby, D. G., Thangaraj, V., Nobre, A. C., Mesulam, M. M., and Warach, S. (1994). Qualitative mapping of cerebral blood flow and functional localization with echo-planar MR imaging and signal targeting with alternating radio frequency. *Radiology*, 192(2), 513–520. doi:10.1148/radiology.192.2.8029425

Elster, D. A. (n.d). MRI Questions & Answers; MR imaging physics & technology. Retrieved from http://mriquestions.com/index.html.

Essig, M., Anzalone, N., Dörfler, A., Enterline, D. S., Law, M., Nguyen, T. B. et al. (2013). Perfusion MRI: The five most frequently asked clinical questions. *AJR American Journal of Roentgenology*, 201(3), W495–W510.

Ewing, J. R. and Bagher-Ebadian, H. (2013). Model selection in measures of vascular parameters using dynamic contrast-enhanced MRI: Experimental and clinical applications. *NMR in Biomedicine*, 26(8), 1028–1041. doi:10.1002/nbm.2996

Golay, X., Hendrikse, J., and Lim, T. C. (2004). Perfusion imaging using arterial spin labeling. *Topics in Magnetic Resonance Imaging: TMRI*, 15(1), 10–27.

Haroon, H. A., Buckley, D. L., Patankar, T. A., Dow, G. R., Rutherford, S. A., Balériaux, D., and Jackson, A. (2004). A comparison of Ktrans measurements obtained with conventional and first pass pharmacokinetic models in human gliomas. *Journal of Magnetic Resonance Imaging*, 19(5), 527–536. doi:10.1002/jmri.20045

Holdsworth, S., and Bammer, R. (2008). Magnetic resonance imaging techniques: fMRI, DWI, and PWI. *Seminars in Neurology*, 28(04), 395–406. doi:10.1055/s-0028-1083697

Hsu, Y. H., Ferl, G. Z., and Ng, C. M. (2013). GPU-accelerated nonparametric kinetic analysis of DCE-MRI data from glioblastoma patients treated with bevacizumab. *Magnetic Resonance Imaging*, 31(4), 618–623. doi:10.1016/j.mri.2012.09.007

Jackson, A., Li, K., and Zhu, X. (2014). Semi-quantitative parameter analysis of DCE-MRI revisited: monte-carlo simulation, clinical comparisons, and clinical validation of measurement errors in patients with type 2 neurofibromatosis. *PLOS One*, 9(3), e90300. doi:10.1371/journal.pone.0090300

Kennan, R. and Jager, H. R. (2003). T2- and T2*-w DCE-MRI: Blood perfusion and volume estimation using bolus tracking. In P. Tofts(Ed.), *Quantitative MRI of the Brain: Measuring Changes Caused by Disease*. Chippenham: John Wiley & Sons. ISBN: 0-470-84721-2

Kim, S. (1995). Quantification of relative cerebral blood flow change by flow-sensitive alternating inversion recovery (FAIR) technique: Application to functional mapping. *Magnetic Resonance in Medicine*, 34(3), 293–301. doi:10.1002/mrm.1910340303

Kim, S., Tsekos, N. V., and Ashe, J. (1997). Multi-slice perfusion-based functional MRI using the FAIR technique: Comparison of CBF and BOLD effects. *NMR in Biomedicine*, 10(4–5), 191–196. doi:10.1002/(sici)1099-1492(199706/08)10:4/5<191::aid-nbm460>3.0.co;2-r

Kiselev, V. (2001). On the theoretical basis of perfusion measurements by dynamic susceptibility contrast MRI. *Magnetic Resonance in Medicine*, 46(6), 1113–1122. doi:10.1002/mrm.1307

Kiselev, V. G. and Posse, S. (1998). Analytical theory of susceptibility induced NMR signal dephasing in a cerebrovascular network. *Physical Review Letters*, 81(25), 5696–5699. doi:10.1103/physrevlett.81.5696

Kiselev, V. and Posse, S. (1999). Analytical model of susceptibility-induced MR signal dephasing: Effect of diffusion in a microvascular network. *Magnetic Resonance in Medicine*, 41(3), 499–509. doi:10.1002/(sici)1522-2594(199903)41:3<499::aid-mrm12>3.3.co;2-f

Larsson, H. B., Courivaud, F., Rostrup, E., and Hansen, A. E. (2009). Measurement of brain perfusion, blood volume, and blood-brain barrier permeability, using dynamic contrast-enhanced T1-weighted MRI at 3 tesla. *Magnetic Resonance in Medicine*, 62(5), 1270–1281. doi:10.1002/mrm.22136

Larsson, H. B, Stubgaard, M., Frederiksen, J. L., et al. (1990). Quantitation of blood–brain barrier defect by magnetic resonance imaging and gadolinium-DTPA in patients with multiple sclerosis and brain tumors. *Magnetic Resonance in Medicine*, 16, 117–131.

Li, K., Zhu, X., Hylton, N., Jahng, G., Weiner, M. W., and Schuff, N. (2005). Four-phase single-capillary stepwise model for kinetics in arterial spin labeling MRI. *Magnetic Resonance in Medicine*, 53(3), 511–518. doi:10.1002/mrm.20390

Lupo, J. M., Lee, M. C., Han, E. T., Cha, S., Chang, S. M., Berger, M. S., and Nelson, S. J. (2006). Feasibility of dynamic susceptibility contrast perfusion MR imaging at 3T using a standard quadrature head coil and eight-channel phased-array coil with and without SENSE reconstruction. *Journal of Magnetic Resonance Imaging*, 24(3), 520–529. doi:10.1002/jmri.20673

Petersen, E. T., Lim, T., and Golay, X. (2006). Model-free arterial spin labeling quantification approach for perfusion MRI. *Magnetic Resonance in Medicine*, 55(2), 219–232. doi:10.1002/mrm.20784

Robson, P. M., Dai, W., Shankaranarayanan, A., Rofsky, N. M., and Alsop, D. C. (2010). Time-resolved vessel-selective digital subtraction MR angiography of the cerebral vasculature with arterial spin labeling. *Radiology*, 257(2), 507–515. doi:10.1148/radiol.10092333

Rosen, B. R., Belliveau, J. W., Aronen, H. J., Kennedy, D., Buchbinder, B. R., Fischman, A. et al. (1991). Susceptibility contrast imaging of cerebral blood volume: Human experience. *Magnetic Resonance in Medicine*, 22(2), 293–299. doi:10.1002/mrm.1910220227

Rosen, B. R., Belliveau, J. W., Brady, T. J., and Vevea, J. M. (1990). Perfusion imaging with NMR contrast agents. *Magnetic Resonance in Medicine*, 14(2), 249–265.

Silva, A. C. and Kim, S. (1999). Pseudo-continuous arterial spin labeling technique for measuring CBF dynamics with high temporal resolution. *Magnetic Resonance in Medicine, 42*(3), 425–429. doi:10.1002/(sici)1522–2594(199909)42:3<425::aid-mrm3>3.3.co;2-j

Simonsen, C. Z., Østergaard, L., Vestergaard-Poulsen, P., Røhl, L., Bjørnerud, A., and Gyldensted, C. (1999). CBF and CBV measurements by USPIO bolus tracking: Reproducibility and comparison with Gd-based values. *Journal of Magnetic Resonance Imaging, 9*(2), 342–347. doi:10.1002/(sici)1522-2586(199902)9:2<342::aid-jmri29>3.3.co;2-2

Sourbron, S. (2010). Technical aspects of MR perfusion. *European Journal of Radiology, 76*(3), 304–313. doi:10.1016/j.ejrad.2010.02.017

Sourbron, S., Ingrisch, M., Siefert, A., Reiser, M., and Herrmann, K. (2009). Quantification of cerebral blood flow, cerebral blood volume, and blood-brain-barrier leakage with DCE-MRI. *Magnetic Resonance in Medicine, 62*(1), 205–217. doi:10.1002/mrm.22005

Sugahara, T., Ikushima, I., Korogi, Y., Kira, T., Liang, L., Shigematsu, Y. et al. (2000). Posttherapeutic intraaxial brain tumor: The value of perfusion-sensitive contrast-enhanced MR imaging for differentiating tumor recurrence from nonneoplastic contrast-enhancing tissue. *AJNR American Journal of Neuroradiology, 21*(5), 901–909.

Tofts, P. S. (1997). Modeling tracer kinetics in dynamic Gd-DTPA MR imaging. *Journal of Magnetic Resonance Imaging, 7*(1), 91–101. doi:10.1002/jmri.1880070113

Tofts, P. S., Brix, G., Buckley, D. L., Evelhoch, J. L., Henderson, E., Knopp, M. V. et al. (1999). Estimating kinetic parameters from dynamic contrast-enhanced t1-weighted MRI of a diffusable tracer: Standardized quantities and symbols. *Journal of Magnetic Resonance Imaging, 10*(3), 223–232. doi:10.1002/(sici)1522-2586(199909)10:3<223::aid-jmri2>3.0.co;2-s

Tofts, P. S. and Kermode, A. G. (1991). Measurement of the blood-brain barrier permeability and leakage space using dynamic MR imaging. 1. Fundamental concepts. *Magnetic Resonance in Medicine, 17*(2), 357–367. doi:10.1002/mrm.1910170208

Wang, J., Alsop, D. C., Detre, J. A., Gonzalez-At, J. B., Li, L., Listerud, J., and Schnall, M. D. (2002). Comparison of quantitative perfusion imaging using arterial spin labeling at 1.5 and 4.0 Tesla. *Magnetic Resonance in Medicine, 48*(2), 242–254.

Welker, K., Boxerman, J., Kalnin, A., Kaufmann, T., Shiroishi, M., and Wintermark, M. (2015). ASFNR recommendations for clinical performance of MR dynamic susceptibility contrast perfusion imaging of the brain. *American Journal of Neuroradiology, 36*(6), E41–E51. doi:10.3174/ajnr.a4341

Williams, D. S., Detre, J. A., Koretsky, A. P., and Leigh, J. S. (1992). Magnetic resonance imaging of perfusion using spin inversion of arterial water. *Proceedings of the National Academy of Sciences of the United States of America, 89*(1), 212–216.

Wong, E. C. (2014). An introduction to ASL labeling techniques. *Journal of Magnetic Resonance Imaging, 40*(1), 1–10. doi:10.1002/jmri.24565

Wong, E. C., Buxton, R. B., and Frank, L. R. (1997). Implementation of quantitative perfusion imaging techniques for functional brain mapping using pulsed arterial spin labeling. *NMR in Biomedicine, 10*(4–5), 237–249. doi:10.1002/(sici)1099-1492(199706/08)10:4/5<237::aid-nbm475>3.0.co;2-x

Wu, W. and Wong, E. C. (2007). Feasibility of velocity selective arterial spin labeling in functional MRI. *Journal of Cerebral Blood Flow & Metabolism, 27*(4), 831–838. doi:10.1038/sj.jcbfm.9600386

Young, G. S. and Qu, J. (2016). Advanced magnetic resonance imaging of brain tumors. In H. B. Newton (Ed.), *Handbook of Neuro-Oncology Neuroimaging* (pp. 167–182). Amsterdam: Elsevier/AP.

Zaharchuk, G. (2007). Theoretical basis of hemodynamic MR imaging techniques to measure cerebral blood volume, cerebral blood flow, and permeability. *American Journal of Neuroradiology, 28*(10), 1850–1858. doi:10.3174/ajnr.a0831

Zaharchuk, G., Ledden, P. J., Kwong, K. K., Reese, T. G., Rosen, B. R., and Wald, L. L. (1999). Multislice perfusion and perfusion territory imaging in humans with separate label and image coils. *Magnetic Resonance in Medicine, 41*(6), 1093–1098. doi:10.1002/(sici)1522-2594(199906)41:6<1093::aid-mrm4>3.0.co;2-0

Zhao, J., Yang, Z., Luo, B., Yang, J., and Chu, J. (2015). Quantitative evaluation of diffusion and dynamic contrast-enhanced MR in tumor parenchyma and peritumoral area for distinction of brain tumors. *PLOS One, 10(9)*. doi:10.1371/journal.pone.0138573

4

Artifacts and Pitfalls of Perfusion MRI

4.1 Introduction

Focus Point

- Subject movement can be a significant source of artifact in any PWI method
- **DSC:**
    - Absolute perfusion quantification is difficult.
    - Bolus dispersion can lead to underestimation of CBF.
    - Blood-brain barrier breakdown can result in contrast leakage.

- **DCE:**
    - EPI susceptibility limitations.
    - Difficulty in the AIF estimation.
    - Systematic errors and variability of the results according to the models used.

- **ASL:**
    - ASL techniques can be prone to static tissue subtraction errors.
    - Quantification model errors.
    - Labeling efficiency issues.

As described in the previous chapter, there are three different MR methods for measuring brain perfusion: the dynamic susceptibility contrast (DSC) and dynamic contrast-enhanced (DCE) imaging methods, which require exogenous contrast administration (intravenous bolus administration of gadolinium), and arterial spin labeling (ASL), which uses the patient's own water molecules as an endogenous diffusible tracer and is performed without exogenous contrast.

Over the past several years, many technical improvements in MR scanner hardware and software (higher field strength, multichannel coils, parallel imaging, etc.) have made these perfusion weighted imaging (PWI) techniques more accurate and robust and have enabled their implementation in research applications and clinical routines.

However, there remain a number of limitations and issues that must be taken into account whenever acquiring and interpreting perfusion methods since there are several assumptions and model fits involved, which can be prone to artifacts and pitfalls.

The scope of this chapter is not to convey the idea that PWI might be unreliable, but on the contrary to introduce and analyze the limitations and pitfalls in a qualitative way, as well as to evaluate possible mitigating strategies or solutions to overcome them. In that sense, the potential user of PWI techniques will be able to avoid or correct errors and make sound interpretations, taking advantage of the very powerful current MR perfusion methodologies, which should be used to provide unique information regarding cerebral hemodynamics.

4.2 Dynamic Susceptibility Contrast (DSC) Imaging Limitations

4.2.1 Subject Motion

As in any other MRI sequence the movement of the patient is a significant source of artifacts. Especially in the DSC technique, due to it involving the rapid injection of a bolus of contrast agent (preferably with an injector system), a noticeable percentage of patients will probably move at the time of the injection. In a small number of patients, this motion can be so severe, that the data cannot be safely processed and the exam is useless.

Fortunately, there is a relatively easy solution if the motion is limited; cut the initial images (before the bolus reaches the brain) during which period the subject is more likely to move due to the injection. The perfusion data can then be calculated using the rest of the images, which can be of good quality. However, there remains the limitation of baseline image calculation since this will not correspond anatomically to the rest of the images, which can substantially degrade the signal–time curves.

A more complete and safer solution is to use image registration techniques to correct for subject movement. First, as a quality control method, an inspection of the raw data before signal processing is recommended so that signal discontinuities due to motion artifacts can be omitted from the baseline estimation or curve integration. Then, image registration for the alignment of the source perfusion dataset with T1-weighted and/or FLAIR images can be used to correct for subject movement (Willats et al., 2006). However, it has to be stressed that while in-plane motion is generally approachable to correct for, through-plane motion correction for 2D acquisitions with interslice spacing is especially difficult (Welker et al., 2015). This is because, in order to increase temporal resolution, limited spatial coverage is usually acquired, resulting in relatively large gaps between the slices acquired (Kosior et al., 2007). Eventually, if the motion exceeds certain limits, no registration technique can overcome the interpretation limitations and the study should be repeated if possible.

4.2.2 Relationship between MR Signal and Contrast Concentration

As discussed in the previous chapter one of the fundamental assumptions used in the DSC modeling approach is the so-called assumption of linearity; that is, the relaxation rate ($R_2 = 1/T_2$) is linearly proportional to the intravascular concentration of the contrast agent (Kiselev, 2001). In fact, based on this assumption, the contrast agent concertation is not directly measured in PWI, but rather indirectly estimated from the changes in the relaxation rates (R_2 or R_2^*). However, it has been shown that this linear relationship may vary, depending on contrast agent concentrations and tissues, with the relaxation rate measured inside arteries varying nonlinearly. Based on theoretical models, it has been shown also that the coupling constant can strongly depend on the pulse sequence employed as well as the vascular morphology. For example, Boxerman et al., (1995) showed that the relaxivity of a tissue will depend on the proportion of different vessels locally involved since there exists a threshold in vessel diameter (10–20 μm) for the relaxivity to change. This would mean that venules and arterioles would contribute differently comparing to capillaries.

Hence the linearity assumption may introduce systematic errors especially regarding absolute quantification (Calamante et al., 2009). The study by Calamante et al. (2009) characterized the associated perfusion errors, concluding that large errors were found when the linear assumption was used and that errors could be greatly reduced when using a quadratic model, and quantifying perfusion as a relative measure. Nevertheless, it has to be stressed that the change in relative parameters (i.e., relative cerebral blood volume [rCBV] and relative mean transit time [MTT]) is less severe.

4.2.3 Bolus Delay and Dispersion

As analytically described in the previous chapter, the perfusion maps used in DSC imaging are calculated from the time curve data using deconvolution of the arterial input function (AIF) (Calamante et al., 2006).

Nevertheless, especially in patients with vascular abnormalities, the bolus can either get delayed in its transit to tissue or be dispersed, resulting in perfusion measurement errors since the AIF is commonly measured in a major artery. Hence, the inaccuracy in the AIF is one of the major potential sources of error in perfusion quantification. Its effect depends on the choice of deconvolution algorithm and has been shown to lead to CBF underestimation and consequently MTT overestimation (Calamante et al., 2000). More specifically, it has been shown that with delays of 1–2 sec, the underestimation in CBF can reach 40% and the overestimation in MTT 60%. Moreover, due to the different delays and dispersions among different areas, this error may vary drastically from one region of the brain to another. One way to mitigate this pitfall is by measuring the AIF locally, as close to the tissue of interest as possible (Lorenz et al., 2006; Willats et al., 2011). Generally, the effect of dispersion is more difficult to mitigate compared to delay since modeling of the vascular effects is required (Calamante et al., 2000; Calamante et al., 2013), and is usually left uncorrected, which should be taken into account, especially in stroke patients. An excellent recent review paper (Calamante, 2013) describes in exhaustive detail all the key factors that should be considered when measuring the AIF in DSC-MRI quantification, and all the issues that users must be aware of when performing or analyzing DSC-MRI.

4.2.4 BBB Disruption and Leakage Correction

Gadolinium-containing contrast agents are normally excluded by the blood-brain barrier (BBB) and cannot enter the extracellular spaces of the brain and spinal cord. The disruption of the BBB caused by diseases like enhancing tumors, subacute infarcts, etc., result in leakage of the contrast agent into the extravascular space. This leakage should be corrected for because it can lead to systematic errors due to the additional pronounced T1- and T2*-relaxation effects that violate the fundamental assumption of tracer kinetic modeling on which DSC is based, that no recirculation of the contrast agent occurs (see Chapter 3, Section 3.2.2). Hence, T1 shortening from extravascular gadolinium can lead to increased signal and may blunt the desired T2* shortening on which DSC is based on.

Several strategies exist to minimize this effect, from simple techniques to more sophisticated models of the first pass kinetics (Bjornerud et al., 2011; Boxerman et al., 2012).

One simple and popular method is called *preloading*. Typically, one-fourth to one-third of the total dose is administered about 5–10 min before the dynamic imaging, and usually another sequence, like a T1-weighted to guide the perfusion section positioning, is run in between. This "preloading" of gadolinium reduces contaminating T1 effects by shortening the pre-bolus intra-voxel T1, raising the baseline signal so that T2* changes can then be better appreciated. Another approach is to reduce the RF-flip angle of the DSC acquisition sequence, although this is substantially restricted by signal-to-noise ratio limitations.

Other methods to compensate for gadolinium extravasation include model-based leakage correction algorithms that linearly fit the $\Delta R2^*(t)$ effect to functions derived from non-enhancing tissue (Quarles et al., 2009; Schmainda et al., 2004), leading to estimates of a leakage contamination term, generating corrected rCBV maps and vascular permeability measurements. Studies have shown that the preloading method with post-processing leakage correction techniques are the two methods that best distinguish rCBV in tumor from normal brain in the presence of gadolimium leakage effects (Paulson and Schmainda, 2008; Welker et al., 2015). In fact, the ASFNR recommendations for clinical performance of MR dynamic susceptibility contrast perfusion imaging of the brain of 2015, highly endorse the use of preload plus model-based post-processing leakage correction for single-echo gadolinium-based DSC-MRI.

4.2.5 Absolute versus Relative Quantification

The most common reported value associated with DSC-MRI techniques in the evaluation of brain tumors is rCBV since it can be derived qualitatively by integrating the relaxivity–time curves without adjustment for the AIF, especially if it is normalized to the contralateral normal appearing white matter (Welker et al., 2015; Sugahara et al., 2000). Thus, a region of perceived perfusion abnormality may be quantitatively evaluated taking into account the contralateral normal region, with the result expressed as a unitless ratio. Another analysis method is the display of histograms, which can provide a meaningful representation of the distribution of perfusion values within a given ROI.

One of the most important points in DSC quantification is the inter- and intra-subject comparisons with contralateral normalization. The main limitation is that measurements may vary with region of interest (ROI) size and more importantly placement, hence the methodology used should be consistent and comparable, avoiding large blood vessels or other structures that may influence the result (Calli et al., 2006). For example, bone or ventricles should always be excluded from a ROI, and a consistent evaluation of the anatomical region is needed since the proportion of gray matter and white matter may also confound perfusion results (Caseiras et al., 2008; Wetzel et al., 2002).

However, this technique of normalization can be subjective and introduces user-dependent variability. Moreover, it can be quite time consuming, especially taking into account the time restrictions during the clinical routine. To mitigate this problem, Bedekar et al., (2010), demonstrated a method called standardization as an objective means of translating all rCBV values to a consistent scale. They stated that this approach can reduce inter-patient and inter-study variability for the same tissue type, thus enabling easy and accurate visual and quantitative comparison across studies (Bedekar et al., 2010).

The subcommittee of the American Society of Functional Neuroradiology (ASFNR) Clinical Practice Committee, in their recommendations for clinical performance of DSC-MRI perfusion imaging (Welker et al., 2015), proposed that "clinical DSC-MRI results should be reported qualitatively unless sufficient age-matched normative data are available for the specific scanner, acquisition technique, and post-processing method employed." It seems that the establishment of an appropriate normative database can be challenging and subjective; hence, relative quantitation may often be more meaningful than absolute quantification. Nevertheless, absolute quantification of CBV and CBF is possible but it requires specialized pulse sequences and acquisition methodologies with adequate processing software in order to derive the AIF (Essig et al., 2013).

Especially for an absolute quantification of CBF (i.e., milliliters per 100 grams per minute), a patient-specific scaling factor should be used. A number of methods have been proposed to mitigate absolute quantification, including ASL measurements deriving a patient-specific correction factor (the ASL- and DSC-CBF ratio) calculated only in short-arrival-time regions (Zaharchuk et al., 2010), and a calibration approach using phase contrast magnetic resonance angiography (PC-MRA) measurements to encode the velocity of flowing spins (Bonekamp et al., 2011).

4.3 Dynamic Contrast Enhancement (DCE) Imaging Limitations

DCE-MRI is generally more complex and challenging than DSC-MRI, especially regarding kinetic analysis and systematic errors that might be introduced. Nevertheless, an increase in the clinical use of DCE-MRI has been noted during the last few years, but mostly in academic medical centers. A convenient and practical approach in brain tumor MR perfusion is to perform DCE-MRI before DSC-MRI so that the administered gadolinium can be used as a preload for leakage correction (see Section 4.2.4).

4.3.1 Suitability of Tumor Lesions

According to the DCE MRI Technical Committee (2012) of the Radiological Society of North America (RSNA), part of the Quantitative Imaging Biomarkers Alliance (QIBA) initiative (http://rsna.org/QIBA_.aspx), there is also the issue of determining the suitability of tumor lesions to perform a DCE exam. "Patients suitable for DCE-analysis must possess at least one tumor >2 cm, away from areas subject to large degrees of cardiac pulsatility artifact, that is not largely cystic or necrotic." Despite the efforts to enroll only suitable patients, on occasion, subsequent analysis might not be feasible due to the following reasons:

- Lack of a tumor of suitable size and/or necrotic
- Failure of gadolinium injection
- Patient motion not correctable with motion-correction algorithms

4.3.2 Subject Motion

Again, subject motion is a significant source of artifacts since various forms of patient motion may contribute to the degradation of signal (respiration, swallowing, involuntary head movement during injection) unless techniques are employed to mitigate their effects. Generally, motion artifacts will largely depend on the choice of imaging sequence used. For example, echo planar imaging (EPI) although very fast can be problematic since it has high susceptibility-related artifacts and is associated with image distortion and "N/2"-type ghosting. Rapid methods employing k-space sub-sampling suffer from point-spread function problems and three-dimensional methods may suffer from phase-wrap effects, and might generally cause a degradation in image quality. Fortunately, motion artifacts in the brain are generally smaller than in other areas of the body, and are amenable to correction via registration strategies as the motion is generally rigid body.

4.3.3 Estimation of Arterial Input Function (AIF)

A fundamental requirement for the extraction of pharmacokinetic information from DCE-MRI data is the accurate determination of the arterial input function (AIF). Although ideally the AIF should be derived from the same DCE-MRI data, in practice, this can be challenging and depends on the location of the measurement or the particular method used to acquire the AIF (Garpebring et al., 2013). In particular, the poor temporal resolution or excitation as well as the flow sensitivity of the method used and saturation effects can lead to degradation of the AIF estimation. Therefore, alternative methods should be employed, like the "population-derived" AIF proposed by Parker et al. (2006), or additional acquisition strategies employing a measurement of a pre-bolus using higher temporal sampling (Viallon et al., 2015).

4.3.4 Temporal and Spatial Resolutions

Temporal and spatial resolution is obviously closely related to the extraction of pharmacokinetic information from DCE-MRI data. Pharmacokinetic modeling requires high temporal resolution to provide a comprehensive description of the underlying causes of contrast agent uptake in a tissue related to the pathology in question (Jahng et al., 2014). Hence very rapid data acquisitions are needed to examine the flow rates in DCE MRI. Unfortunately, there is a price tag for this and that is low signal to noise ratio (SNR) or loss of precision. Both temporal and spatial resolution may affect the estimation of AIF as mentioned above. Low temporal resolution may lead to inaccurate AIF while low spatial resolution may lead to a partial volume effect and introduce inaccuracies in AIF estimation.

4.3.5 Variability of Results According to the Models Used

The accuracy and precision of the derived DCE parameters is vital before it can be used on a larger scale since the aforementioned variety of factors may affect the clinical information. Many efforts are currently under way to evaluate the accuracy and assess the reliability of different models and software, using clinical data as well as phantoms (Huang et al., 2014; Kim et al., 2014; Kudo et al., 2013; Shin et al., 2014). For example in the work by Huang et al. twelve (12) different software packages were compared on the same DCE datasets and considerable differences were reported. Nevertheless, despite the considerable differences, nearly all algorithms managed a satisfactory prediction of the therapy response, which suggests that DCE has the potential to be used in clinical routines if the consistency in the acquisition and data post-processing and evaluation is optimized and standardized.

A major limitation that has to be taken into account is the relatively high temporal resolution needed for the models used (about 1 sec, while a typical temporal resolution currently used is 5 sec). Therefore, their use is still relatively confined. Another limitation is that multiple models may be appropriate for a given tissue type and this is a typical problem associated with DCE data post-processing. Obviously, the results would then depend on the chosen model (Haroon et al., 2004; Sourbron, 2010). Therefore an appropriate selection is a crucial parameter.

The easiest approach to tackle the model selection problem or a "general rule of thumb" is the following: the most appropriate model is the simplest that provides a good fit to the data (Sourbron, 2010)!

Nevertheless, the more complex the tissue behavior, the more important the fitting difficulties of the models due to inhomogeneities since identical effects are modeled in undifferentiated ways, causing kinetic time-course inaccuracies (Buckley, 2002). Hence, the quantification model should be improved to a so-called "tissue homogeneity model," to consider the incomplete exchange of intravascular and extravascular water.

Conversely, a "compensation" of the model for effects that have been omitted using less detailed parameters may introduce systematic errors into the hemodynamic modeling, again causing inaccuracies. In conclusion, there is always a tradeoff between precision and accuracy, as fitting processes become more stable (better and more robust models) and noise levels are reduced.

4.3.6 Quality Assurance

Another issue in DCE-MRI is that a number of scanner dependent factors may unfavorably affect the quality of derived data. These factors include the stability of the MR scanner over the duration of the DCE-MRI protocol, the accuracy of the predetermined T1 relaxation estimates (derived from the static and dynamic data), geometric distortion over a large field of view, and the excitation slice profile (http://rsna.org/QIBA_.aspx). Additionally, there can be lack of consistency in patient set-up, difficulty in matching and image fusing anatomical locations over sequential studies, and challenges in ROI definition and placement for analysis. Quality assessment and control procedures should be an essential part of the DCE protocols, and should be tested and validated prior to the commencement of the exam. The stability of the scanner should be evaluated using dedicated test objects as well as volunteers for the estimation of artifacts, especially regarding the T1 measurements.

In conclusion, the quality of derived DCE-MRI data can be optimized, provided appropriate quality control procedures are followed and staff are appropriately trained to recognize the problems that may occur.

4.4 Arterial Spin Labeling (ASL) Imaging Limitations

There are a number of artifacts and pitfalls that can potentially affect correct perfusion quantification in the ASL techniques. The main potential sources of errors are described in the following section.

4.4.1 Subject Motion

Patient motion is a common problem and a significant source of artifacts in ASL. The main reason is that, due to the relatively low SNR of the perfusion-weighted images in ASL, multiple repetitions are required (~40) to acquire enough data to increase it to satisfactory levels, resulting in very long scan times (>30 min), making the sequence sensitive to motion artifacts. In particular, patient movement between control and tagging sequences may appear as peripheral blurring around the area of interest or as variations in brightness, as illustrated in Figure 4.1.

Despite the use of signal intensity averaging and spatial coregistration steps, the most consistent motion-related pattern observed is high signal intensity strips in the periphery of the brain, producing false hypo- or hyper-perfusion signals (Deibler et al., 2008). Moreover, when there are large position changes or sudden movements, image registration approaches may be of limited efficacy and it might be preferable to exclude the involved images from the data postprocessing with little decrease in SNR (Ferré et al., 2013).

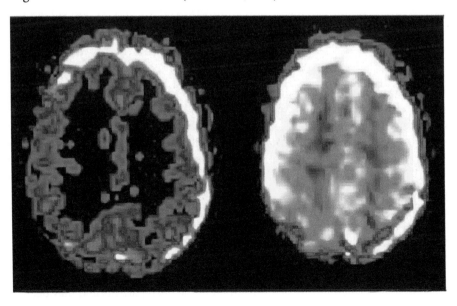

FIGURE 4.1 Motion artifacts on successive ASL slices manifest by varying contrast and a bright halo at the edge of the images. (Courtesy of Allen D. Elster, MRIquestions.com.)

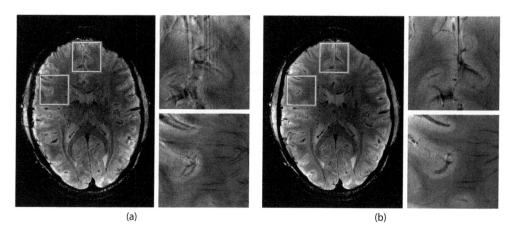

(a)   (b)

FIGURE 4.2   A slice by slice comparison of the corrected and uncorrected gradient echo images reveals the superior quality of the data acquired with motion correction. The images acquired without motion correction (a) show significant motion artifacts and blurring compared to (b). (From Stucht, D. et al., *PLOS One*, 10, e0133921, 2015. With permission.)

Very recently, conventional motion correction approaches have been replaced by image-based prospective motion correction techniques, promising to alleviate the effects of even large position changes during the scan (Aksoy et al., 2014; Zun et al., 2014). One such method of image-based prospective motion correction is called PROMO (**PRO**spective **MO**tion correction) and uses three orthogonal 2D navigator images with rigid-body tracking algorithm to correct for patient motion in real time. An example of the superior quality of motion correction algorithms is illustrated in Figure 4.2.

A slice by slice comparison of the corrected and uncorrected gradient echo images from Stucht et al. (2015) reveals the superior quality of the data acquired with motion correction.

State-of-the-art implementation of ASL includes background suppression with a separate neck labeling coil, which has been shown to improve sensitivity and reproducibility (Shen and Duong, 2011).

4.4.2 Physiological Signal Variations

Not infrequently, physiological modifications of signal intensity are observed and should be recognized in order to avoid mistaking them for pathology. If patients are scanned with their eyes open, bilateral regional increases in signal intensity of the occipital lobes can be observed, corresponding to visual cortex activation (Deibler et al., 2008). Amukotuwa et al. (2016) reported that following visual cortex activation, an increased occipital lobe CBF is more frequent and conspicuous with 2D PASL (utilizing an EPI readout) than with 3D pCASL, although they could not explain why (Amukotuwa et al., 2016). They overall suggest that it is preferable to ask patients to keep their eyes closed during the ASL sequence since this physiological occipital hyperperfusion is difficult to control for in the clinical population and may eventually mask or distract from pathological changes in CBF. Cortex activation has also been involved with phenomena related to the transit time and residual vascular signal (Ferré et al., 2013). Generally, a hyperfrontal pattern of regional CBF distribution has been previously reported using several perfusion methods; it is believed to be a normal finding in young and middle-aged patients and may decrease both with normal aging and with increasing cerebrovascular risk factors (Deibler et al., 2008).

Another physiological variation is an increased signal-to-noise ratio as well as globally elevated absolute CBF values in pediatric patients. In fact, CBF values have been reported to increase to peak levels at about 3–8 years, and then gradually decrease to adult levels (Epstein, 1999; Wang et al., 2003). Deibler et al. report that possible explanations for this globally increased perfusion include higher baseline CBF, faster mean transit time, increased baseline magnetization values in gray and white matter, and increased T1 values in blood and tissue. Moreover, the immature paranasal sinus development in pediatric patients may play a role in the increased signal due to the decreased susceptibility artifacts.

On the other hand, one can observe areas of physiological hypoperfusion. These areas extend from the frontal and occipital horns to the frontal and parieto-occipital cortex. It is assumed that they generally correspond to the cerebral border zones between distal end branches of intracranial arteries (Hendrikse et al., 2008). This aspect may be important as it has been associated with the difficulty of obtaining reliable CBF measurements in the white matter (van Osch et al., 2009).

4.4.3 Magnetic Susceptibility Artifacts

The base of the skull, the temporal and frontal region as well as air field cavities are vulnerable to magnetic susceptibility artifacts, particularly visible with rapid acquisition EPI sequences, and are represented as signal intensity voids on ASL CBF maps. Furthermore, susceptibility artifacts are present around metallic implants and early post-surgical materials. Hematomas and calcifications might also be a problem in ASL as they may mask the underlying high signal intensity on ASL CBF maps. (Ferré et al., 2013). More specifically, calcified masses (e.g., meningiomas or oligodendrogliomas) may be particularly problematic in the assessment of tumor vascularity since susceptibility and neovascularity may coexist, producing competing effects (Deibler et al., 2008). These artifacts can be mitigated by improving imaging quality, using less sensitive image acquisition methods, parallel imaging and multi-channel coils (Boss et al., 2007; Ferré et al., 2012). However, in some instances ASL susceptibility artifacts manifest differently than conventional images. For example, in Figure 4.3a susceptibility artifact easily recognized on the source image may mimic a region of hypoperfusion/infarct on the ASL image.

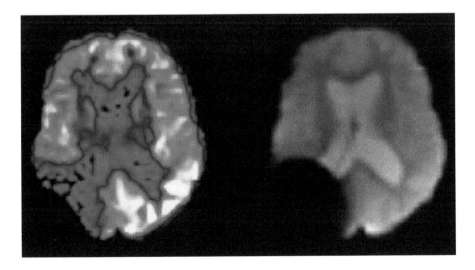

FIGURE 4.3 A susceptibility artifact easily recognized on the source image (right) that mimics a region of hypoperfusion/infarct on the ASL image (left). (Courtesy of Allen D. Elster, MRIquestions.com.)

4.4.4 Coil Sensitivity Variations

It is possible that the receiver coils may exhibit asymmetric sensitivity and spatial variations due to defective elements. This might lead to regional signal differentiations adjacent to the problematic element, mimicking areas of hyper- or hypoperfusion on ASL images. This artifact can be detected by evaluating the M0 image. It is then necessary to confirm and mitigate it, either by using an alternative coil if available, or by repositioning the patient's head in a different area of the coil and repeating the exam. It can then be eliminated by image filtering or other post-processing techniques. Future applications of ASL may include M0 mapping and application of weighting factors to compensate for coil sensitivity issues (Deibler et al., 2008).

4.4.5 Labeling Efficiency

One of the most important parameters in ASL is obviously the labeling process. Since the overall signal is based on the subtraction of a "labeled" image (or "label" image) from an "unlabeled" image (or "reference" image) efficient labeling is the key to success. It is evident that labeling errors will produce errors in the perfusion parameters estimation. The labeling efficiency obtained in practice differs substantially between PASL and CASL sequences. For example, in the CASL technique, the labeling efficiency (also known as degree of inversion) is about 70% (Alsop and Detre, 1998). On the other hand, efficiency close to the ideal of 100% is often obtained using adiabatic RF pulses in PASL (Frank et al., 1997). Hardware limitations and power deposition restrictions may further reduce efficiency (Utting et al., 2003).

Another limitation that needs to be taken into account is that sections of the brain acquired toward the end of the labeling volume (please refer to Figure 3.9) may be less inversed (i.e., contain less label) than those acquired at the beginning. Consequently, signal intensity may be reduced in the more rostral images of the CBF maps when images are acquired from inferior to superior (Deibler et al., 2008).

To improve the overall labeling efficiency, there was a continuous effort by many research groups to further improve the labeling pulses. These efforts include, but are not limited to; mitigation of less labeling using a constant delay time (Yongbi et al., 2002), hyperbolic scant pulse (Silver et al., 1985), frequency offset corrected inversion pulse (Ordidge et al., 1996), and bandwidth-modulated adiabatic selective saturation and inversion pulse (Warnking and Pike, 2004).

4.4.6 Transit Time Effects

Another major pitfall in the quantitative estimate of cerebral perfusion with ASL is an intrinsic sensitivity of the method to the determination of the arterial transit time (ATT), that is, the time difference between the application of the spin inversion (or the tag) and the arrival of the tagged volume. The problem is that the ATT differs across the brain, even in healthy subjects, and that the ASL signal is exponentially dependent on this value. It may also vary regionally inside the brain as well as according to pathologies leading to under- or overestimation of the CBF values (Qiu et al., 2010). This effect is more important in CASL techniques due to the much longer distance the blood has to travel between the labeling and exchange sites (Calamante et al., 2009).

Several techniques have been proposed to mitigate transit time effects, like the quantitative imaging of perfusion using a single subtraction (Alsop and Detre, 1996; Wong et al., 1998) or the concept of multiple post-labeling delay (PLD) acquisitions (Günther et al., 2001). Transit time is generally reduced with CASL and pCASL, while in PASL saturation pulses can be applied to the tagged volume to bypass the problem (Viallon et al., 2015).

In CASL transit time estimation typically requires the acquisition of multiple images with different delay time, while in PASL, different labeling time is used (Chen et al., 2012). It should be noted that the measurement of transit time involves the acquisition of a considerable amount of data, and therefore can be very time consuming. Moreover, both approaches cause an extension of scan time and this is a disadvantage as the risk for subject movement increases.

4.4.7 Errors from Quantification Models

There are several sources of error in the quantification of ASL perfusion values and therefore the models used should be as accurate and detailed as possible in order not to introduce systematic errors in the quantification procedure. The factors that affect quantification can either be assumed or directly measured and include the ATT, the T1 values of blood and tissue, the labeling efficiency and capillary permeability.

Most models include assumptions for these factors, which will be a source of error in perfusion quantification. For example, (1) most models use a fixed value for the T1 of blood since its estimation is difficult due to movement; (2) most models use the assumption of the free diffusion of water in a well-mixed tissue compartment although multiple compartments is a more accurate approach (Jahng et al., 2014); (3) many models may not include a vascular volume V_p, which can lead to significant errors in the fitted values of K^{trans} (Buckley, 2002).

Nevertheless, in order for quantification of perfusion to be valid, the mitigation of errors should be implemented as follows:

1. Tissue T1 should be measured although this can lead to an increase of the total acquisition time. A very quick approach for T1 mapping based on two inversion recovery images and a reference image manages to map tissue T1 within a few seconds (Jahng et al., 2005, 2014).
2. To mitigate the error introduced from the single compartment models, separate blood and tissue component models should be preferred (Alsop and Detre, 1996).

4.5 Conclusions and Future Perspectives

As discussed in this chapter, the accuracy of brain perfusion quantification can be affected by a number of factors depending on the technique used. However, the scope of this chapter was not to convey the idea that perfusion techniques might be unreliable, but on the contrary to introduce and analyze the potential problems in terms of imaging artifacts and processing-related factors. Thus, through increased knowledge, these limitations can be overcome, and allow their integration into modern clinical routine protocols. Indeed, these are powerful methodologies that should be used to provide unique information regarding cerebral hemodynamics through tumor cellularity, tissue invasion, metabolism, and microvasculature. Nevertheless, many of the artifacts shown in this chapter can be eliminated or significantly improved if promptly recognized, while solutions to many of these issues are the subject of current research. It seems that in the future, the use of protocols that combine exogenous contrast methods (DSC and DCE) with the endogenous contrast in ASL will have a great impact in the clinical routine.

References

Aksoy, M., Maclaren, J. R., Ooi, M., et al. (2014). Prospective optical motion correction for 3D pseudo-continuous arterial spin labeling. *Proceedings of the 22nd Annual Meeting ISMRM*, Milan, Italy, 424.

Alsop, D. C. and Detre, J. A. (1996). Reduced transit-time sensitivity in noninvasive magnetic resonance imaging of human cerebral blood flow. *Journal of Cerebral Blood Flow & Metabolism, 16*(6), 1236–1249. doi:10.1097/00004647-199611000-00019

Alsop, D. C. and Detre, J. A. (1998). Multisection cerebral blood flow MR imaging with continuous arterial spin labeling. *Radiology, 208*(2), 410–416. doi:10.1148/radiology.208.2.9680569

Amukotuwa, S. A., Yu, C., and Zaharchuk, G. (2016). 3D Pseudocontinuous arterial spin labeling in routine clinical practice: A review of clinically significant artifacts. *Journal of Magnetic Resonance Imaging, 43*(1), 11–27. doi:10.1002/jmri.24873

Bedekar, D., Jensen, T., and Schmainda, K. M. (2010). Standardization of relative cerebral blood volume (rCBV) image maps for ease of both inter- and intrapatient comparisons. *Magnetic Resonance in Medicine, 64*(3), 907–913. doi:10.1002/mrm.22445

Bjornerud, A., Sorensen, A. G., Mouridsen, K., and Emblem, K. E. (2011). T1- and T*2-dominant extravasation correction in DSC-MRI: Part I—Theoretical considerations and implications for assessment of tumor hemodynamic properties. *Journal of Cerebral Blood Flow & Metabolism, 31*(10), 2041–2053. doi:10.1038/jcbfm.2011.52

Bonekamp, D., Degaonkar, M., and Barker, P. B. (2011). Quantitative cerebral blood flow in dynamic susceptibility contrast MRI using total cerebral flow from phase contrast magnetic resonance angiography. *Magnetic Resonance in Medicine, 66*(1), 57–66. doi:10.1002/mrm.22776

Boss, A., Martirosian, P., Klose, U., Nägele, T., Claussen, C. D., and Schick, F. (2007). FAIR-TrueFISP imaging of cerebral perfusion in areas of high magnetic susceptibility differences at 1.5 and 3 Tesla. *Journal of Magnetic Resonance Imaging, 25*(5), 924–931. doi:10.1002/jmri.20893

Boxerman, J. L., Hamberg, L. M., Rosen, B. R., and Weisskoff, R. M. (1995). MR contrast due to intravascular magnetic susceptibility perturbations. *Magnetic Resonance in Medicine, 34*(4), 555–566. doi:10.1002/mrm.1910340412

Boxerman, J. L., Prah, D. E., Paulson, E. S., Machan, J. T., Bedekar, D., and Schmainda, K. M. (2012). The Role of preload and leakage correction in gadolinium-based cerebral blood volume estimation determined by comparison with MION as a criterion standard. *American Journal of Neuroradiology, 33*(6), 1081–1087. doi:10.3174/ajnr.a2934

Buckley, D. L. (2002). Uncertainty in the analysis of tracer kinetics using dynamic contrast-enhanced T1-weighted MRI. *Magnetic Resonance in Medicine, 47*(3), 601–606.

Calamante, F. (2013). Arterial input function in perfusion MRI: A comprehensive review. *Progress in Nuclear Magnetic Resonance Spectroscopy, 74*, 1–32. doi:10.1016/j.pnmrs.2013.04.002

Calamante, F., Connelly, A., and Osch, M. J. (2009). Nonlinear ΔR^*2 effects in perfusion quantification using bolus-tracking MRI. *Magnetic Resonance in Medicine, 61*(2), 486–492. doi:10.1002/mrm.21839

Calamante, F., Gadian, D. G., and Connelly, A. (2000). Delay and dispersion effects in dynamic susceptibility contrast MRI: Simulations using singular value decomposition. *Magnetic Resonance in Medicine, 44*(3), 466–473. doi:10.1002/1522-2594(200009)44:3<466::aid-mrm18>3.3.co;2-d

Calamante, F., Willats, L., Gadian, D. G., and Connelly, A. (2006). Bolus delay and dispersion in perfusion MRI: Implications for tissue predictor models in stroke. *Magnetic Resonance in Medicine, 55*(5), 1180–1185. doi:10.1002/mrm.20873

Calli, C., Kitis, O., Yunten, N., Yurtseven, T., Islekel, S., and Akalin, T. (2006). Perfusion and diffusion MR imaging in enhancing malignant cerebral tumors. *European Journal of Radiology, 58*(3), 394–403. doi:10.1016/j.ejrad.2005.12.032

Caseiras, G. B., Thornton, J., Yousry, T., Benton, C., Rees, J., Waldman, A., and Jager, H. (2008). Inclusion or exclusion of intratumoral vessels in relative cerebral blood volume characterization in low-grade gliomas: Does it make a difference? *American Journal of Neuroradiology, 29*(6), 1140–1141. doi:10.3174/ajnr.a0993

Chen, Y., Wang, D. J., and Detre, J. A. (2012). Comparison of arterial transit times estimated using arterial spin labeling. *Magnetic Resonance Materials in Physics, Biology and Medicine, 25*(2), 135–144. doi:10.1007/s10334-011-0276-5

DCE MRI Technical Committee. (2012, July). DCE MRI quantification profile, quantitative imaging biomarkers alliance. Version 1.0. Reviewed Draft. *QIBA.* Retrieved from http://rsna.org/QIBA_.aspx.

Deibler, A., Pollock, J., Kraft, R., Tan, H., Burdette, J., and Maldjian, J. (2008). Arterial spin-labeling in routine clinical practice, part 1: Technique and artifacts. *American Journal of Neuroradiology, 29*(7), 1228–1234. doi:10.3174/ajnr.a1030.

Epstein, H. T. (1999). Stages of increased cerebral blood flow accompany stages of rapid brain growth. *Brain and Development, 21*(8), 535–539. doi:10.1016/s0387-7604(99)00066-2

Essig, M., Shiroishi, M. S., Nguyen, T. B., Saake, M., Provenzale, J. M., Enterline, D. et al. (2013). Perfusion MRI: The five most frequently asked technical questions. *American Journal of Roentgenology, 200*(1), 24–34. doi:10.2214/ajr.12.9543

Ferré, J., Bannier, E., Carsin-Nicol, B., Gauvrit, J., Mineur, G., and Raoult, H. (2013). Arterial spin labeling (ASL) perfusion: Techniques and clinical use. *Diagnostic and Interventional Imaging, 94*(12), 1211–1223.

Ferré, J., Petr, J., Bannier, E., Barillot, C., and Gauvrit, J. (2012). Improving quality of arterial spin labeling MR imaging at 3 tesla with a 32-channel coil and parallel imaging. *Journal of Magnetic Resonance Imaging, 35*(5), 1233–1239. doi:10.1002/jmri.23586

Frank, L. R., Wong, E. C., and Buxton, R. B. (1997). Slice profile effects in adiabatic inversion: Application to multislice perfusion imaging. *Magnetic Resonance in Medicine, 38*(4), 558–564. doi:10.1002/mrm.1910380409

Garpebring, A., Brynolfsson, P., Yu, J., Wirestam, R., Johansson, A., Asklund, T., and Karlsson, M. (2013). Uncertainty estimation in dynamic contrast-enhanced MRI. *Magnetic Resonance in Medicine, 69*(4), 992–1002. doi:10.1002/mrm.24328

Günther, M., Bock, M., and Schad, L. R. (2001). Arterial spin labeling in combination with a look-locker sampling strategy: Inflow turbo-sampling EPI-FAIR (ITS-FAIR). *Magnetic Resonance in Medicine, 46*(5), 974–984. doi:10.1002/mrm.1284

Haroon, H. A., Buckley, D. L., Patankar, T. A., Dow, G. R., Rutherford, S. A., Balériaux, D., and Jackson, A. (2004). A comparison of Ktrans measurements obtained with conventional and first pass pharmacokinetic models in human gliomas. *Journal of Magnetic Resonance Imaging, 19*(5), 527–536. doi:10.1002/jmri.20045

Hendrikse, J., Petersen, E. T., Laar, P. J., and Golay, X. (2008). Cerebral border zones between distal end branches of intracranial arteries: MR Imaging. *Radiology, 246*(2), 572–580. doi:10.1148/radiol.2461062100

Huang, W., Barbodiak, D. P., Chen, Y., Chang, M., Chenevert, T. L., Dyvorne, H. et al. (2014). Variations of dynamic contrast-enhanced magnetic resonance imaging in evaluation of breast cancer therapy response: A multicenter data analysis challenge. *Translational Oncology, 7*(1), 153–166.

Jahng, G., Calamante, F., Li, K., and Ostergaard, L. (2014). Perfusion magnetic resonance imaging: A comprehensive update on principles and techniques. *Korean Journal of Radiology, 15*, 554–577.

Jahng, G., Stables, L., Ebel, A., Matson, G. B., Meyerhoff, D. J., Weiner, M. W., and Schuff, N. (2005). Sensitive and fast T1 mapping based on two inversion recovery images and a reference image. *Medical Physics, 32*(6 Part 1), 1524–1528. doi:10.1118/1.1915014

Kim, H. S., Goh, M. J., Kim, N., Choi, C. G., Kim, S. J., and Kim, J. H. (2014). Which combination of MR imaging modalities is best for predicting recurrent glioblastoma? Study of diagnostic accuracy and reproducibility. *Radiology, 273*(3), 831–843. doi:10.1148/radiol.14132868

Kiselev, V. (2001). On the theoretical basis of perfusion measurements by dynamic susceptibility contrast MRI. *Magnetic Resonance in Medicine, 46*(6), 1113–1122. doi:10.1002/mrm.1307

Kosior, R. K., Kosior, J. C., and Frayne, R. (2007). Improved dynamic susceptibility contrast (DSC)-MR perfusion estimates by motion correction. *Journal of Magnetic Resonance Imaging, 26*(4), 1167–1172. doi:10.1002/jmri.21128

Kudo, K., Christensen, S., Sasaki, M., Østergaard, L., Shirato, H., Ogasawara, K. et al. (2013). Accuracy and reliability assessment of CT and MR perfusion analysis software using a digital phantom. *Radiology, 267*(1), 201–211. doi:10.1148/radiol.12112618.

Lorenz, C., Benner, T., Chen, P. J., Lopez, C. J., Ay, H., Zhu, M. et al. (2006). Automated perfusion-weighted MRI using localized arterial input functions. *Journal of Magnetic Resonance Imaging, 24*(5), 1133–1139. doi:10.1002/jmri.20717

Ordidge, R. J., Wylezinska, M., Hugg, J. W., Butterworth, E., and Franconi, F. (1996). Frequency off-set corrected inversion (FOCI) pulses for use in localized spectroscopy. *Magnetic Resonance in Medicine, 36*(4), 562–566. doi:10.1002/mrm.1910360410

Parker, G. J., Roberts, C., Macdonald, A., Buonaccorsi, G. A., Cheung, S., Buckley, D. L. et al. (2006). Experimentally-derived functional form for a population-averaged high-temporal-resolution arterial input function for dynamic contrast-enhanced MRI. *Magnetic Resonance in Medicine, 56*(5), 993–1000. doi:10.1002/mrm.21066

Paulson, E. S. and Schmainda, K. M. (2008). Comparison of dynamic susceptibility-weighted contrast-enhanced MR methods: Recommendations for measuring relative cerebral blood volume in brain tumors. *Radiology, 249*(2), 601–613. doi:10.1148/radiol.2492071659

Qiu, M., Maguire, R. P., Arora, J., Planeta-Wilson, B., Weinzimmer, D., Wang, J. et al. (2010). Arterial transit time effects in pulsed arterial spin labeling CBF mapping: Insight from a PET and MR study in normal human subjects. *Magnetic Resonance in Medicine, 63*(2), 374–384. doi:10.1002/mrm.22218

Quarles, C. C., Gochberg, D. F., Gore, J. C., and Yankeelov, T. E. (2009). A theoretical framework to model DSC-MRI data acquired in the presence of contrast agent extravasation. *Physics in Medicine and Biology, 54*(19), 5749–5766. doi:10.1088/0031-9155/54/19/006

Schmainda, K. M., Badruddoja, M. A., Joseph, A. M., Krouwer, H. G., Lund, R., Pathak, A. P. et al. (2004). Characterization of a first-pass gradient-echo spin-echo method to predict brain tumor grade and angiogenesis. AJNR. *American Journal of Neuroradiology, 25*(9), 1524–1532.

Shen, Q. and Duong, T. Q. (2011). Background suppression in arterial spin labeling MRI with a separate neck labeling coil. *NMR in Biomedicine, 24*(9), 1111–1118. doi:10.1002/nbm.1666

Shin, K., Ahn, K., Choi, H., Jung, S., Kim, B., Jeon, S., and Hong, Y. (2014). DCE and DSC MR perfusion imaging in the differentiation of recurrent tumour from treatment-related changes in patients with glioma. *Clinical Radiology, 69*(6), e264–e272. doi:10.1016/j.crad.2014.01.016

Silver, M. S., Joseph, R. I., and Hoult, D. I. (1985). Selective spin inversion in nuclear magnetic resonance and coherent optics through an exact solution of the Bloch-Riccati equation. *Physical Review A, 31*(4), 2753–2755. doi:10.1103/physreva.31.2753

Sourbron, S. (2010). Technical aspects of MR perfusion. *European Journal of Radiology, 76*(3), 304–313. doi:10.1016/j.ejrad.2010.02.017

Stucht, D., Danishad, K. A., Schulze, P., Godenschweger, F., Zaitsev, M., and Speck, O. (2015). Highest resolution in vivo human brain MRI using prospective motion correction. *PLOS One, 10*(7), e0133921. doi:10.1371/journal.pone.0133921

Sugahara, T., Ikushima, I., Korogi, Y., Kira, T., Liang, L., Shigematsu, Y. et al. (2000). Posttherapeutic intraaxial brain tumor: The value of perfusion-sensitive contrast-enhanced MR imaging for differentiating tumor recurrence from nonneoplastic contrast-enhancing tissue. AJNR. *American Journal of Neuroradiology, 21*(5), 901–909.

Utting, J. F., Thomas, D. L., Gadian, D. G., and Ordidge, R. J. (2003). Velocity-driven adiabatic fast passage for arterial spin labeling: Results from a computer model. *Magnetic Resonance in Medicine, 49*(2), 398–401. doi:10.1002/mrm.10363

van Osch, M. J., Buchem, M. A., Hendrikse, J., Kies, D. A., Teeuwisse, W. M., and Walderveen, M. A. (2009). Can arterial spin labeling detect white matter perfusion signal? *Magnetic Resonance in Medicine, 62*(1), 165–173.

Viallon, M., Cuvinciuc, V., Delattre, B., Merlini, L., Barnaure-Nachbar, I., Toso-Patel, S. et al. (2015). Erratum to: State-of-the-art MRI techniques in neuroradiology: Principles, pitfalls, and clinical applications. *Neuroradiology, 57*(10), 1075–1075. doi:10.1007/s00234-015-1548-y

Wang, J., Detre, J. A., Haselgrove, J., Jahng, G., Licht, D. J., Liu, C. et al. (2003). Pediatric perfusion imaging using pulsed arterial spin labeling. *Journal of Magnetic Resonance Imaging: JMRI, 18*(4), 404–413.

Warnking, J. M. and Pike, G. B. (2004). Bandwidth-modulated adiabatic RF pulses for uniform selective saturation and inversion. *Magnetic Resonance in Medicine, 52*(5), 1190–1199. doi:10.1002/mrm.20262

Welker, K., Boxerman, J., Kalnin, A., Kaufmann, T., Shiroishi, M., and Wintermark, M. (2015). ASFNR recommendations for clinical performance of MR dynamic susceptibility contrast perfusion imaging of the brain. *American Journal of Neuroradiology, 36*(6), E41–E51. doi:10.3174/ajnr.a4341

Wetzel, S. G., Cha, S., Johnson, G., Lee, P., Law, M., Kasow, D. L. et al. (2002). Relative cerebral blood volume measurements in intracranial mass lesions: Interobserver and intraobserver reproducibility study. *Radiology, 224*(3), 797–803. doi:10.1148/radiol.2243011014

Willats, L., Christensen, S., Ma, H. K., Donnan, G. A., Connelly, A., and Calamante, F. (2011). Validating a local Arterial Input Function method for improved perfusion quantification in stroke. *Journal of Cerebral Blood Flow & Metabolism, 31*(11), 2189–2198. doi:10.1038/jcbfm.2011.78

Willats, L., Connelly, A., and Calamante, F. (2006). Improved deconvolution of perfusion MRI data in the presence of bolus delay and dispersion. *Magnetic Resonance in Medicine, 56*(1), 146–156. doi:10.1002/mrm.20940

Wong, E. C., Buxton, R. B., and Frank, L. R. (1998). Quantitative imaging of perfusion using a single subtraction (QUIPSS and QUIPSS II). *Magnetic Resonance in Medicine, 39*(5), 702–708. doi:10.1002/mrm.1910390506

Yongbi, M. N., Fera, F., Yang, Y., Frank, J. A., and Duyn, J. H. (2002). Pulsed arterial spin labeling: Comparison of multisection baseline and functional MR imaging perfusion signal at 1.5 and 3.0 T: Initial results in six subjects. *Radiology, 222*(2), 569–575. doi:10.1148/radiol.2222001697

Zaharchuk, G., Straka, M., Marks, M. P., Albers, G. W., Moseley, M. E., and Bammer, R. (2010). Combined arterial spin label and dynamic susceptibility contrast measurement of cerebral blood flow. *Magnetic Resonance in Medicine, 63*(6), 1548–1556. doi:10.1002/mrm.22329

Zun, Z., Shankaranarayanan, A., and Zaharchuk, G. (2014). Pseudocontinuous arterial spin labeling with prospective motion correction (PCASL-PROMO). *Magnetic Resonance in Medicine, 72*(4), 1049–1056. doi:10.1002/mrm.25024

5

Magnetic Resonance Spectroscopy

5.1 Introduction

Focus Point

- MRS can be considered as a bridge between the anatomic and physiological information and the metabolic characteristics of tissue in vivo.
- The principal phenomenon of MRS is the "chemical shift," which is directly related to the biochemical environment of every nucleus.
- The proton nucleus is the most useful nucleus for MRS, due to its high natural abundance (>99.9%) and intrinsic sensitivity (high gyromagnetic ratio γ).

Magnetic resonance spectroscopy (MRS) is a technique which provides a non-invasive method for characterizing the cellular biochemistry of brain pathologies, as well as for monitoring the biochemical changes after treatment in vivo. In that sense, it can be considered a bridge between the anatomical and physiological information and the metabolic characteristics of tissue in vivo (Soares and Law, 2009).

The principal phenomenon is the so-called "chemical shift," which is caused by the unique (for every nucleus) shielding from the external magnetic field (B_0) by the electrons surrounding them. Hence, this chemical shift effect is directly related to the biochemical environment of the nuclei. The electron magnetic moment opposes the primary applied magnetic field B_0; therefore, the more the electrons the less the magnetic field the nuclei will "feel." This "feeling" can be expressed as the effective magnetic field B_n of the nucleus:

$$B_n = B_0 (1-\sigma) \tag{5.1}$$

where σ, is the screening constant, which is proportional to the chemical environment of the nucleus. Hence, in vivo MRS is a combination of MR imaging and chemical spectroscopy, where instead of an image, a spectrum of resonant peaks is produced.

Due to the chemical shift phenomenon, it is evident that MRS is feasible on any nucleus possessing a magnetic moment, such as a proton (^1H), carbon-13 (^{13}C), phosphorus (^{31}P), and sodium (^{23}Na).

Early MRS studies were focused on the phosphorus nucleus (^{31}P) since this was the most technically feasible at the early 1980s when in vivo MRS became possible (Luyten et al., 1989). In recent years, though, proton MRS (^1H-MRS) has become much more popular as it is possible to obtain high resolution spectra in reasonably short scan times (Frahm et al., 1989; Soares and Law, 2009) due to proton's high natural abundance (>99.9%) and intrinsic sensitivity (high gyromagnetic ratio γ). Until now, ¹H-MRS has been used both as a research as well as a clinical tool for detecting abnormalities, visible or not yet visible, on conventional MRI. Suggestively, Möller-Hartman et al. reported that when only the MR images were used for radiological diagnosis of focal intracranial mass lesions, their type and grade were correctly identified in 55% of the cases. The addition of MR spectroscopic information significantly increased the proportion of correctly diagnosed cases to 71% (Möller-Hartmann et al., 2002).

Figure 5.1 illustrates typical examples of magnetic resonance spectra of a 62-year old male with a glioma from (a) the lesion and (b) the contralateral normal parenchyma. The detection

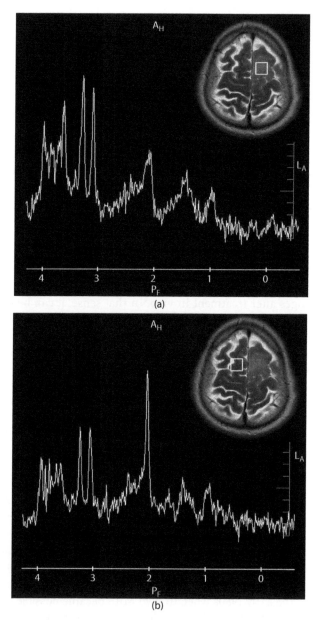

FIGURE 5.1 Typical example of single voxel magnetic resonance spectroscopy of a 62-year-old male with a glioma. (a) Spectrum from the lesion. (b) Spectrum from the contralateral normal parenchyma.

of spatial or signal abnormalities as a result of the disease conditions is evident by pure visual evaluation of the spectra.

The application of ¹H-MRS has been always challenging in terms of its technical requisites (field homogeneity, gradients, coils and software), as well as the accurate metabolic interpretation with regard to pathological processes. Despite the challenges, the clinical applications of ¹H-MRS are continuously increasing as the clinical hardware has become more robust and user-friendly along with improved data analysis, spectra post-processing techniques and metabolite interpretation confidence.

The success of MRS as a valuable clinical tool depends on the accuracy of the acquired data as well as correct post-process and analyses. The purpose of this chapter is to elaborately introduce the current status of ¹H-MRS in terms of the metabolites detected in the brain with their clinical usefulness, including the technical considerations, the acquisition, and post-processing methods.

5.2 MRS Basic Principles Explained

Focus Point

- The position of the peaks denotes frequency and determines certain metabolite presence based on their chemical shift.
- The area under the peak is roughly proportional to the concentration of metabolites.
- The ppm unit represents frequencies as a fraction of their absolute resonance frequency, and is independent of field strength.

Proton, derived from the Greek word πρῶτον (meaning "first") is a subatomic particle with a positive electric charge and one-half spin, and exhibits the electromagnetic properties of a dipole magnet. This name was given to the hydrogen nucleus by Ernest Rutherford in 1920.

When protons are placed in an external magnetic field B_0, they align themselves along the direction of the field (either parallel or anti-parallel) and demonstrate a circular oscillation. The frequency of this circular motion (called Larmor frequency) is dependent on the strength of the local magnetic field and the molecular structures to which protons belong. This can be expressed by the Larmor equation:

$$\omega_0 = \gamma B_0 \tag{5.2}$$

where ω_0 is the Larmor frequency, γ is the gyromagnetic ratio specific for the nucleus, and B_0 is the strength of the external magnetic field.

When a RF pulse (in other words electromagnetic energy) is supplied at this frequency, the molecules absorb this energy and change their alignment. When the RF pulse is switched off, the molecules realign themselves to the magnetic field by releasing their absorbed energy. This released energy is the basis of the MR signal and hence MR imaging.

Proton MRS, or ¹H-MRS, uses the same hardware as conventional MRI, however, the main difference is that the frequency of the MR signal is used to encode different types of information. MRI generates structural images, whereas ¹H-MRS provides chemical information about the tissue under study. The MRS technique also uses gradients to selectively excite a particular volume of tissue (a so-called voxel), but rather than creating an image of it, it records the free induction decay (FID) and produces a spectrum from that voxel.

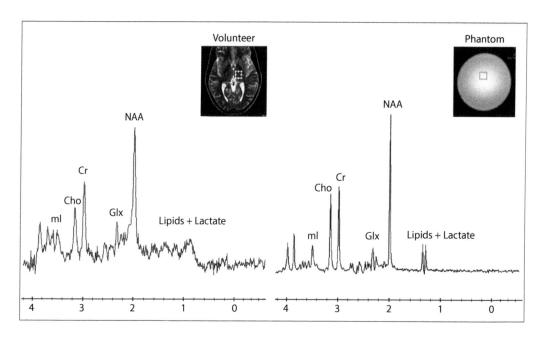

FIGURE 5.2 Left: ^{1}H-MRS spectrum from the white matter (WM) measured at 3T in the brain of a 19-year-old healthy volunteer. Right: Spectrum from a standard spectroscopy phantom (25-cm-diameter MRS HD sphere; General Electric Company). The metabolites in the phantom are 3.0 mmol/L choline chloride, 10.0 mmol/L creatine hydrate, 12.5 mmol/L N-acetylaspartic acid, 7.5 mmol/L myo-inositol, 12.5 mmol/L L-glutamic acid, and 5 mmol/L lactate, containing 0.1% sodium azide, 0.1% Magnavis, 50 mmol/L potassium dihydrogen phosphate, and 56 mmol/L sodium hydroxide. PRESS, TE = 35 ms, TR = 1500 ms, voxel size 20 × 20 × 20 mm³, 128 signal averages.

Figure 5.2 illustrates a ^{1}H-MRS spectrum from the white matter (WM) of a 19-year-old healthy volunteer measured at 3T (left) and (right) a spectrum from a standard spectroscopy phantom (25-cm-diameter MRS HD sphere; General Electric Company).

The vertical axis (*y*) represents the signal intensity or relative concentration for the various cerebral metabolites and the horizontal axis (*x*) represents the chemical shift frequency in parts per million (ppm). ppm is commonly used instead of frequency in Hz because the ppm unit represents frequencies as a fraction of their absolute resonance frequency, and is independent of field strength. Hence spectra originating from different magnet strengths (e.g., 1.5T vs. 3T) can be directly compared.

The nature of the chemical shift effect is to produce a change in the resonant frequency for nuclei of the same type attached to different chemical species. The phenomenon is due to variations in surrounding electron clouds of neighboring atoms, which shield nuclei from the main magnetic field (B_0). The resulting frequency difference can be used to identify the presence of important chemical compounds.

In other words, metabolites can be differentiated based on their slightly different resonant frequencies due to their different local chemical environments. This separation (i.e., the chemical shift) is depicted as a ppm difference on the horizontal axis of spectra.

The only pre-requirement is that the water peak must be suppressed so that the relatively low concentration metabolites (about four orders of magnitude lower) can be evaluated.

In Figure 5.2 it can be noticed that 0 ppm is on the right-hand side and that the *x*-axis limit is 4 ppm. That is because above 4 ppm the suppression of the water peak (more specifically at 4.7 ppm) makes the neighboring portions of the spectrum unreliable. It must also be noted that the phantom spectrum of Figure 5.2 (right) has been corrected for temperature. The signal is inversely proportional to the absolute temperature of the tissue or the object under evaluation (Kreis, 1997). As the temperature is reduced, the Boltzmann distribution gives a larger difference between spin populations, and hence the magnetization of the sample increases (Hoult and Richards, 1976).Hence, the phantom temperature, which is about 20° Celsius (at scanner room temperature) causes a shift in the spectrum of about 0.1 ppm to the right and needs to be corrected for in order to compare spectra.

Within the spectrum, metabolites, due to the chemical shift effect, are characterized by one or more peaks with a certain resonance frequency, line width (full width at half maximum of the peak's height, FWHM), line shape (e.g., Lorentzian or Gaussian), phase, and peak area according to the number of protons that contribute to the observed signal. By monitoring those peaks, ^{1}H-MRS can provide a qualitative and/or a quantitative (provided there is adequate signal post-processing) analysis of a number of metabolites within the brain if a reference of known metabolite concentration is used at a particular field strength (Christiansen et al., 1993; Jansen et al., 2006; Sarchielli et al., 1999; Sibtain et al., 2007). Generally, the relative areas under each peak are roughly proportional to the number of nuclei in that particular chemical environment.

5.2.1 Technical Issues

Focus Point

- Increased field strength (B0):
  - Increased B0 increases SNR and chemical shift leading to improved spectral resolution and better visualization of the weakly represented neurochemicals.
  - Comes with a price tag: increased spatial misregistration and magnetic susceptibility of the paramagnetic materials.

- Shimming:
  - The process of optimizing the magnetic field homogeneity over the ROI. Efficient shimming results in improved spectral resolution.

- Voxel positioning:
  - Cautious spatial localization of the voxel removes unwanted signals from outside the ROI and avoids partial volume effects.

- Voxel size:
  - A practical minimum for the voxel size in in-vivo MRS is **1 cm^3**.

5.2.2 Data Acquisition

In order to acquire high quality spectra, several technical considerations should be taken into account concerning the available MRS techniques, the applied magnetic field, the shimming procedures, as well as the good control of the spatial origin of the spectra.

Spectra can be acquired either with a single voxel (SV) technique (single voxel spectroscopy, SVS) or multiple voxel technique, known as either magnetic resonance spectroscopic imaging (MRSI) or chemical shift imaging (CSI) in two (2D CSI) or three dimensions (3D CSI). Of course, the more voxels (2D) or slices (3D), the more time is needed for the acquisition, and the greater the possibility for subject movement. Figure 5.3 schematically illustrates the three MRS techniques.

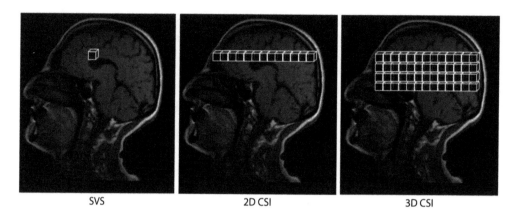

| SVS | 2D CSI | 3D CSI |

FIGURE 5.3   Spectra can be acquired either with a single voxel technique (single voxel spectroscopy, SVS) or multiple voxel technique, alternatively called chemical shift imaging (CSI), either in 2D or 3D.

SVS is based on the point resolved spectroscopy (PRESS) (Bottomley and Park, 1984) or the stimulated echo acquisition mode (STEAM) (Frahm et al., 1989) pulse sequences while MRSI uses a variety of pulse sequences (Spin Echo, PRESS, etc.) (Brown et al., 1982; Duyn and Moonen, 1994; Luyten et al., 1990).

PRESS uses a 90° and two 180° RF pulses in a fashion similar to a standard multi-echo sequence (Figure 5.4b). Each RF pulse is applied using a different physical gradient as the slice selection gradient. Only protons located at the point where all three pulses intersect produce the spin echo at the desired TE.

STEAM uses three selective 90° pulses, each with a gradient on one of the three axes as shown in Figure 5.4a, and is designed to collect only the stimulated echo signal from the area (voxel) of the intersection of all three slices. STEAM has been used for many studies because for many years it was the only sequence capable of short echo times (~30 ms) (Bottomley and Park, 1984; Frahm et al., 1989).

Inevitably the question arises: Which pulse sequence should be used?

The answer is that although there are differences between the two, these are rather subtle and that in practice this mainly depends on the particular availability from the scanner vendor. Nevertheless, there has been a detailed comparison from Moonen et al. (1989) who concluded that the major difference is in the nature of the echo signal. In PRESS, the entire net magnetization from the voxel is refocused to produce the echo signal, whereas in STEAM, a maximum of one-half of the entire net magnetization generates the stimulated echo. As a result, PRESS has a SNR significantly larger than for STEAM for equivalent scan parameters. Another difference is that the voxel dimensions with PRESS may be limited by the high transmitter power for the 180° RF pulses. Finally, and more importantly, STEAM allows for shorter TE values, reducing signal losses from T2 relaxation and allowing observation of metabolites with short $T2^*$.

Although SVS is very useful in clinical practice for several reasons (widely available, short scan times, short TE contains signals from more compounds, etc.), its main disadvantage is that it does not address spatial heterogeneity of spectral patterns and in the context of brain

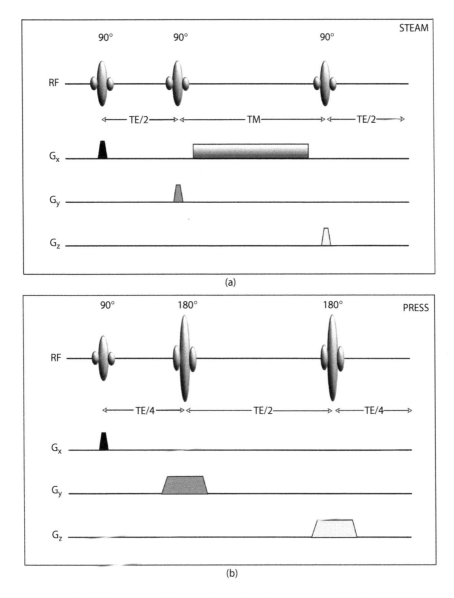

FIGURE 5.4 Simplified diagrams of the single voxel pulse sequences. (a) the STEAM sequence and (b) the PRESS sequence.

tumors. These factors are particularly important, especially for treatment planning in case of radiation or surgical resection.

Hence, a lesion's heterogeneity is better assessed by MRSI. MRSI techniques have been extended to two dimensions (2D) by using phase-encoding gradients in two directions (Duyn et al., 1993; Luyten et al., 1990), or, subsequently, three-dimensional (3D) encoding (Gruber et al., 2003; Nelson et al., 1999). Thus, MRSI techniques allow the detection of localized ^{1}H-MR spectra from a multidimensional array of locations (see Figure 5.5). While technically more challenging due to (1) possibility for significant magnetic field inhomogeneity across the entire volume of interest, (2) the so called "voxel bleeding," which is a spectral degradation due to intervoxel contamination, (3) longer data acquisition times and (4) challenging post-processing of large

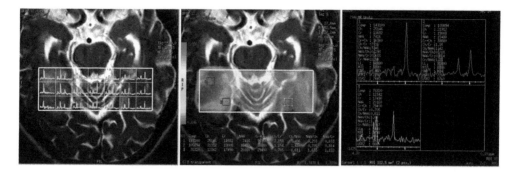

FIGURE 5.5   Example of a 2D-MRSI of a patient with a glioma. Left: The lesion on a T2 weighted image. Middle: MRSI data presented as a metabolic map of Choline/Creatine with voxels from multiple locations at the same plane of the lesion. Right: Multiple spectra with metabolite ratios.

multidimensional datasets, MRSI can detect metabolic profiles from multiple spatial positions, thereby offering an unbiased characterization of the entire object under investigation.

> Again, inevitably the question arises: Which one should be used in clinical practice, SVS, CSI, or both?

This is not an easy answer. Usually, SVS is performed at short TEs (35 ms) while MRSI at long TEs (135–144 ms). As already mentioned, short TE spectra contain signals from more compounds and have better SNRs; however, they have worse water and lipid contamination when compared to long TEs. On the other hand, long TE spectra have lower SNR and fewer detectable compounds, but they are better resolved with flatter baselines and contain more information in considerably less time. Moreover, MRSI can produce metabolic maps and therefore it can reveal abnormalities in multiple locations. Consequently, the method of choice depends on the clinical information required as well as the position of the area under investigation. If for example a lesion is very close to areas with high magnetic susceptibility (e.g., the sinuses) then MRSI may be impossible to acquire. On the other hand, if spectroscopy is being used to evaluate a disease that is diffuse or covers a large area of anatomy, then spectra from several volumes of tissue can be measured simultaneously, which is advantageous and time saving. Not infrequently, or maybe rather most commonly, a combination of both techniques proves to be the best solution.

5.2.3 Field Strength (B_0)

From the very beginning of MRI, optimum field strength was a topic of debate. Nevertheless, for MRS applications the substantial benefit from a high magnetic field was already known for a long time in ex vivo NMR and animal studies. In contrast to MRI, in ^{1}H-MRS clinical applications, a magnetic field of sufficient strength is preferable as it is not the signals of water and fat that are of interest, but rather the considerably smaller metabolites' signals (about four orders of magnitude). Therefore, most clinical ^{1}H-MRS measurements are performed using MR systems with field strengths of >1.5T. As we all know now, the brilliance of high-field strengths and especially 3T won the race, although even more powerful (4T, 7T, and even higher) body scanners are currently in use.

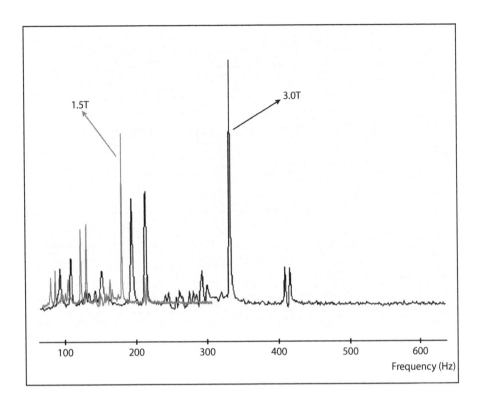

FIGURE 5.6 Comparison of single voxel spectra obtained at 1.5T and 3T, using the protocol (PROBE-P), a standard spectroscopy phantom (25-cm-diameter MRS HD sphere; General Electric Company) and the software provided by the manufacturer (General Electric Medical Systems, Milwaukee, WI). At 3T, the signal-to-noise ratio (SNR) is about 25% higher and the spectral distance between the metabolites (in Hertz) is doubled.

Obviously, the main advantage of increasing the magnetic field strength is the subsequent increase of the signal-to-noise ratio (SNR) since the intensity of the MR signal is correlated linearly with the strength of the static magnetic field. Thus, theoretically, the signal-to-noise ratio (SNR) would double when doubling the field strength (e.g., from 1.5T to 3T); however, when put into clinical practice, the improvement ranges only from 20% to 50%. This is very well illustrated in Figure 5.6 where the overlapping of spectra indicatively emphasizes the differences between 1.5T and 3.0T. It is shown that at 3T, the SNR is about 25% higher and the spectral distance between the metabolites (in Hertz) is approximately doubled.

In the study by Barker et al. (2001), a 28% increase in SNR at 3T compared to that of 1.5T at short TEs was demonstrated, appreciably less than the theoretical 100% improvement. The limited SNR improvement can be explained by several factors including T1 and T2 relaxation, coil and system losses, and RF penetration effects, which strongly affect the linearity between SNR and field strength, as well as type of sequence, number of signal averages and size of sample volume (Di Costanzo et al., 2007; Edelstein et al., 1986; Ocali et al., 1998). Nevertheless, by using particular methods of data acquisition, processing and fitting, SNR can increase by about 80% at 4T compared with that at 1.5T (Bartha et al., 2000) and approximately 100% at 7T compared with 4 T (Tkác et al., 2001). One of the possible approaches to increase the SNR is the use of multiple receiver coils. In fact, a well-designed phased array (PA) head coil has significantly superior sensitivity to that of the birdcage-type volume coil, which is more widely used (De Zwart et al., 2004).

Another advantage of the higher magnetic field, is the proportional increase of the Chemical Shift, from 220 Hz at 1.5T to 440 Hz at 3T (Alvarez-Linera, 2008). Consequently, this is reflected by a more effective water suppression and improved baseline separation of J-coupled metabolites, without the need of sophisticated spectral editing techniques (Barker et al., 2001; Bartha et al., 2000; Stephenson et al., 2011). The improvement in spectral resolution is more evident at 7T where weakly represented neurochemicals with important clinical impact, such as scyllo-Ins, aspartate, taurine and NAAG, can be clearly estimated (Stephenson et al., 2011; Tkác et al., 2001).

5.2.4 Voxel Size Dependency

At lower field strengths (≤1.5 T) the suggested minimum voxel size is $2 \times 2 \times 2$ cm (i.e., 8 cm³). At higher field strengths (≥3 T) most ¹H-MRS studies have been performed with a minimum spatial resolution of 1 cm³ (Gruber et al., 2003). Figure 5.7 depicts the dependency of SNR to the voxel size. It is evident that increasing the voxel sizes increases the SNR while the acquisition time remains constant.

Reducing voxel's size substantially reduces the SNR since these two are linearly proportional. Nevertheless, at fields of 3T or higher, voxels below 1 cm³ can be obtained with acceptable SNR (Boer et al., 2011; Gruber et al., 2003), but with rather long acquisition times (see also Section 6.2.4). Hence, reduced voxel sizes can improve the sensitivity and specificity of diagnosis and enable the creation of metabolic maps that depict details of heterogeneous lesions such as gliomas, where changes in their development can occur in small areas. Nevertheless, intrinsic field-dependent technical difficulties may affect the aforementioned advantages and should be considered. When the frequency shift between two adjacent nuclei is large enough, a measurable alteration of the MR signal, used to encode the *x*- and *y*-axis spatial coordinates, will occur producing a spatial misregistration. This means that the volume of MRS information may not be the same as that displayed on the localizer MR image (Di Costanzo et al., 2003). More importantly, in high magnetic field strengths, magnetic susceptibility from paramagnetic substances and blood products are sensibly increased (Gu et al., 2002). Consequently, magnetic field inhomogeneity and susceptibility artifacts complicate the extraction of good-quality spectra, especially from largely heterogeneous lesions

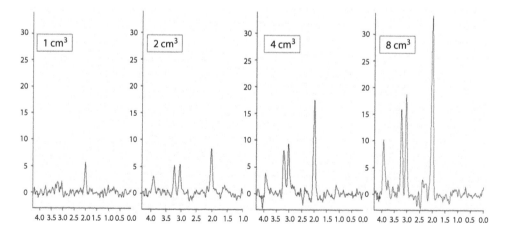

FIGURE 5.7   Dependency of SNR to the voxel size. Spatial resolution of 1 cm³ is practically considered a minimum.

(Di Costanzo et al., 2003). Nevertheless, improvement of the local shimming methods can alleviate the problem (see next section).

5.2.5 Shimming

Shimming refers to the process of adjusting the field gradients of the scanner system in order to optimize the magnetic field homogeneity over the volume under study.

In case you are wondering (as I did), the word *shimming* originates from the thin sheets of iron that were originally used by engineers to adjust the uniformity of magnets called *"shims."* Magnetic field inhomogeneities result primarily from susceptibility differences between different tissues and between tissue and air cavities, which are scaled non-linearly in ultra-high magnetic fields (Avdievich et al., 2009). Thus, voxels that are placed in, or near inhomogeneous regions of the brain, such as the temporal poles, are difficult to shim due to their close proximity to the sinuses.

In any case it has to be stressed here that there is no such thing as a perfectly homogeneous magnetic field because it cannot be produced, and even if it could, the insertion of the patient into the scanner would immediately make it imperfect due to the so-called susceptibility effects. These effects are produced by inhomogeneities such as the aforementioned air-filled areas (sinuses or intestines) or dense bone (skull base), which change the main magnetic field in their immediate vicinity, so tissues around such inhomogeneities will experience different magnetic fields.

Hence, field homogeneity is specified by measuring the full width at half maximum (FWHM) of the water resonance, which determines the spectral resolution. Special emphasis must be given to shimming, especially in high field strengths, as it increases both sensitivity and spectral resolution. This is why most devices come equipped with second or third order shimming (Avdievich et al., 2009) by monitoring either the time domain or frequency domain of the ^{1}H-MRS signal (Drost et al., 2002). Especially in cases when field homogeneity should be reached in large volumes of interest, as in the case of MRSI, 4-order shimming might be necessary (Hetherington et al., 2006; Gillard et al., 2010. Figure 5.8 illustrates the difference between a good shimming and a bad shimming procedure.

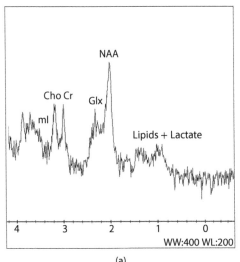 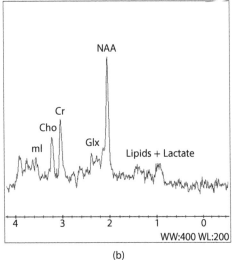

(a) (b)

FIGURE 5.8 Notice the difference between (a) inadequate shimming (linewidth = 13 Hz) and (b) adequate shimming (linewidth = 5 Hz).

A minimum of 6 Hz at 1.5T and 10 Hz at 3.0T linewidth is considered to be an adequate shimming for SVS, taking into account that 0.1 ppm equals approximately to 6 Hz at 1.5T. Shimming can be performed either manually or automatically. Manual shimming requires a great deal of expertise and therefore, in a well-maintained scanner, automatic shimming is considered a safer and more reproducible option.

Effective shimming requires methods for mapping a field's strength variations over the area under study. Methods that have been developed for field mapping can be grouped into two categories: those based on 3D field mapping (Hetherington et al., 2006; Miyasaka et al., 2006) and those that map the magnetic field along projections (Zhang et al., 2009). In both shimming methods, information about the magnetic field variation is calculated from phase differences acquired during the evolution of the magnetization in a non-homogeneous field.

5.2.6 Water and Lipid Suppression Techniques

Water protons are at such a high concentration compared with the other proton containing metabolites, the water peak obviously dominates the spectrum and makes the other peaks impossible to measure. The same applies for lipids, hence water and peri-cranial lipid suppression techniques are of paramount importance in ¹H-MRS procedures in order to observe in detail the much less concentrated metabolite signals.

The metabolites of interest are usually several orders of magnitude less in concentration than water. Therefore, the water suppression efficiency should be robust and should not vary spatially across the field of view (FOV). This is illustrated in Figure 5.9.

The existing water suppression techniques can be divided into three major groups, namely: (1) methods that employ frequency-selective excitation and/or refocusing pulses; or (2) utilize differences in relaxation parameters; and (3) other methods, including software-based water suppression.

The most common method of the first group utilizes multiple (typically three) frequency-selective 90° pulses (chemical shift selective water suppression (CHESS) pulses (Haase et al., 1985), prior to localization pulse sequence. Additionally, suppression can be achieved

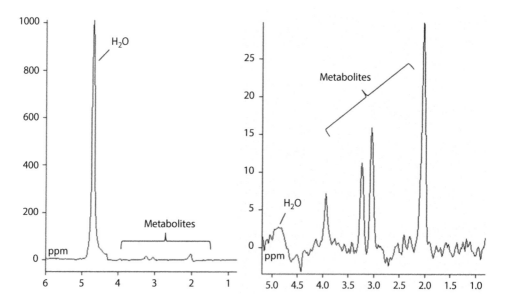

FIGURE 5.9   Left: Before water suppression, the metabolite peaks under investigation are depicted as noise. Right: After water suppression, the metabolites are revealed.

by selectively diphase water, while metabolites of interest are rephased using refocusing pulses during the spin-echo period. The three spin-echo-based methods are WATERGATE (Piotto et al., 1992), excitation sculpting (Hwang and Shaka, 1995) and MEGA (Mescher et al., 1996).

As water and metabolite T1s are sufficiently different, it is possible to suppress the water signal and observe the metabolites in the close proximity to the water resonance. WEFT (water eliminated Fourier transform) is among the oldest T1-based water suppression methods (Patt and Sykes, 1972), which is identical to an inversion recovery sequence used for T1 relaxation time measurements. Water suppression can be further optimized by applying a band-selective inversion with a gradient dephasing pulse (BASING), alone or with addition of CHESS (Star-Lack et al., 1997).

The third method includes two separate scans with metabolite resonances inversion. The water resonance is not inverted in either scan. Therefore, the difference between the first and second scan will result in the subtraction of water and hence water suppressed metabolite spectrum.

Lipid suppression can be performed by avoiding the excitation of the lipid signal using STEAM or PRESS localization, hence suppressing their contribution to the detected signal.

Opposite to the strategy employed by volume pre-localization, outer volume suppression pulses (OVS) are applied to pre-saturate the lipid signal (Duyn et al., 1993). As illustrated in Figure 5.10, rather than avoiding the spatial selection of lipids, OVS excites narrow slices of the brain's lipid-rich regions.

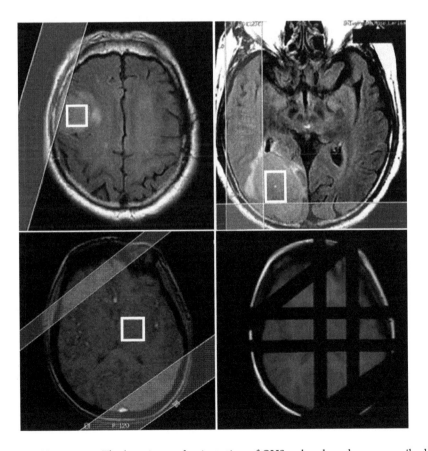

FIGURE 5.10 Upper part: The location and orientation of OVS pulses have been prescribed in order to saturate as much peri-cranial lipid as possible while the signal within the voxel remains unperturbed. Lower part, left: The application of two extra saturation bands. Lower part, right: The two extra bands with the default bands are illustrated.

Additionally, the difference in T1s of lipids (250–350 ms) and metabolites (1000–2000 ms) allow the application of an inversion pulse (inversion time ~200 ms), which will selectively null the lipid signal (Spielman et al., 1992). If the inversion delay is appropriately chosen, the longitudinal lipid magnetization will be zero, so the lipids will be nulled. The aforementioned suppression methods are usually better implemented in long TE spectra as both water and lipid resonances have shorter T2 relaxation times than many metabolites.

At this point it is very important to stress that MRS sequences apply a number of pre-saturation bands by default (left-right, posterior-anterior, superior-inferior) as illustrated in Figure 5.10, lower right. The addition of extra saturation bands is possible but there is always a limit (usually 10, i.e., +4). After this limit, the extra bands gradually delete default saturation bands with an unknown order and this has to be taken into account!

5.3 MRS Metabolites and Their Biological and Clinical Significance

Focus Point

- Metabolite ratios provide robust in vivo markers of biochemistry but they should be interpreted with caution.
- N-acetyl-aspartate: Detected in neuronal and non-neuronal cells. Considered as a neuron's integrity index.
- Total choline: Metabolic index of membrane density and integrity.
- Total creatine: Present in both neuronal and glial cells. Serves as an energy buffer and energy shuttle. Can be used as the reference metabolite.
- Myo-inositol: Glial and myelin degradation marker. Malignant tumor indicator.
- Lactate: Present in both intracellular and extracellular spaces. Product of anaerobic glycolysis. Increases in tumors.
- Lipids: Membrane breakdown and necrosis index. Increases in high grade tumors.
- Glx complex (Glu/Gln): Provides detoxification and regulation of neurotransmitters. Difficult to separate complex's peaks.

The aim of in-vivo ^{1}H-MRS is to use metabolite levels for an accurate evaluation of cerebral lesions, and therefore, determination of the relationship between metabolic profile and pathologic processes is required. The biological and clinical significance of the basic resonances in a spectrum, as well as the less commonly detected ones, is discussed in this section. For a more comprehensive overview of the assignment and significant role of each brain metabolite evaluated, a summary is depicted in Table 5.1. The light shade represents difficulty in separating low filed strengths, and the darker shade represents the metabolites that are elevated/detected only in pathological conditions. Basic resonances detected in the normal brain are not shaded.

5.3.1 Myo-Inositol

Starting from left to right of the indicative spectrum of Figure 5.2a, Myo-inositol (mI–*3.56 ppm*) is a cyclic sugar alcohol that gives rise to four groups of resonances with the

TABLE 5.1 List of Metabolites Detected in the Human Brain by Proton MR Spectroscopy

Metabolite	Chemical Shift	Physiological Significance	Increased	Decreased
NAA (N-acetyl aspartate, other N-acetyl moieties)	2.02 ppm	Healthy neuronal cell marker. Only seen in nervous tissue. Exact physiological role uncertain.	Very rarely Canavan's disease	Commonly: nonspecific neuronal loss or dysfunction due to range of insults, including ischemia, trauma, gliosis inflammation, infection, tumors, dementia,
Cho Choline containing compounds	3.2 ppm	Detectable resonance is predominantly choline derivatives. Marker of membrane turnover. Higher in WM than GM. Increase with age.	Tumors, inflammation, chronic hypoxia	Stroke, encephalopathy (hepatic human immunodeficiency virus (HIV)/liver disease
Cr Creatine/ Phospho-creatine	3.0 ppm	Compounds related to energy storage; thought to be marker of energetic status of cells. Other metabolites are frequently expressed as ratio to Cr. Low in infants. Increases with age.	Trauma, hyperosmolar states	Hypoxia, stroke, tumors
mI(Myo) Myo-inositol (other inositols)	3.56 ppm (short TE only)	Pentose sugar, osmolyte, glial cell marker. High in infants.	Neonates, Alzheimer's, diabetes, low-grade glioma recovered encephalopathy, hyperosmolar states	Malignant tumors, chronic hepatic encephalopathy, stroke
Glx Glutamate (Glu) + Glutamine (Gln)	2.1–2.4 ppm (short TE only)	Complex overlapping J–coupled resonances difficult to separate and quantify at clinical field strengths (1.5–3T). Amino acid neurotransmitters Glu excitatory, Gln inhibitory.	Hepatic encephalopathy, severe hypoxia, OTC deficiency	Possibly Alzheimer's disease
Lactate	1.35 ppm (doublet, 7 ppm separation)	Not seen in normal adult brain. End product of anaerobic respiration. Thought to be elevated in foamymacrophages.	Ischemia, inborn errors of metabolism tumors (all grades), abscesses, inflammation	-
Lipids Mobile lipid moieties	0.9 and 1.3 ppm (short TE unless ↑↑)	Not seen in normal brain. Membrane breakdown/ lipid droplet formation. May precede histological necrosis.	High-grade tumors, abscesses, acute inflammation, acute stroke	-

(Continued)

TABLE 5.1 (*Continued*)   List of Metabolites Detected in the Human Brain by Proton MR Spectroscopy

Metabolite	Chemical Shift	Physiological Significance	Increased	Decreased
Alanine	1.47 ppm	Amino acid present in the normal brain.	Meningiomas (?)	
Succinate, acetate, amino acids	2.4 ppm, 1.92 ppm, 0.9 ppm	Products of bacterial metabolism.	Pyogenic abscesses	-
Acetoacetate, acetone	Not normally detectable	Only pathologically elevated in inborn errors.	Inborn errors of metabolism	-
Mannitol, ethanol	Various	Administered drugs and other substances.		-

larger and most important signal occurring at 3.56 ppm. It is observable on short time echo (TE) spectra as it exhibits short T2 relaxation times and is susceptible to dephasing effects due to J-coupling.

Interestingly, the exact function of mI is still uncertain (Ross, 1991), however it has been proposed as a glial marker (Brand et al., 1993) and an increase of mI levels is believed to represent glial proliferation or an increase in glial cell size, both of which may occur in inflammation (Soares and Law, 2009). Additionally, this metabolite is involved in the activation of protein C kinase, which leads to production of proteolytic enzymes found in malignant and aggressive cerebral tumors, serving as a possible index for glioma grading (Castillo et al., 2000). mI has also been labeled as a breakdown product of myelin (Kruse et al., 1993) since altered levels of mI have been encountered in patients with degenerative and demyelinating diseases (Lin et al., 2005; Wang et al., 2009; Wattjes and Barkhof, 2008; Wattjes et al., 2009).

5.3.2 Choline-Containing Compounds

Choline-containing compounds (tCho–*3.22 ppm*) comprise signals from free choline (Cho), phosphocholine (PC) and glycerophosphocholine (GPC), with a resonant peak located at 3.22 ppm. Since the resonance contains contributions from several methyl proton choline-containing compounds, it is often referred as "total Choline" (tCho).

tCho is involved in pathways of phospholipid synthesis and degradation, thus reflecting a metabolic index of membrane density and integrity as well as membrane turnover (Howe et al., 2003; Möller-Hartmann et al., 2002; Nelson et al., 1999; Sibtain et al., 2007; Soares and Law, 2009).

Consistent changes of tCho signals have been observed in a large number of cerebral diseases. Processes that lead to elevation of tCho include accelerated membrane synthesis of rapidly dividing cancer cells in brain tumors (Howe et al., 2003; Sibtain et al., 2007; Kararizou et al., 2006; Soares and Law, 2009), cerebral infractions, infectious diseases (Calli et al., 2002; Lai et al., 2008), and inflammatory-demyelinating diseases (Hayashi et al., 2003; Malhotra et al., 2009).

It has to be stressed that tCho exhibits a physiologic marked regional variability with higher concentrations observed in the pons and lower levels in the vermis and dentate (Mascalchi et al., 2002). Detailed knowledge about regional variations of tCho is necessary for an accurate interpretation of tCho levels, especially in diseases such as epilepsy and psychiatric disorders where tCho levels are subtly different to normal levels.

5.3.3 Creatine and Phosphocreatine

Creatine and Phosphocreatine (tCr–*3.03 ppm and 3.93 ppm*) together are often referred as total creatine (tCr) because they cannot be distinguished with a standard clinical MR unit (up to 7T) and their sum is thus mentioned. Cr and PCr arise from the methyl and methylene protons of Cr and phosphorylated Cr. tCr is located at 3.03 ppm and 3.93 ppm resonant frequencies.

In the brain tCr is present in both neuronal and glial cells and is involved in energy metabolism serving as an energy buffer via the creatine kinase reaction retaining constant ATP levels; it also serves as an energy shuttle, diffusing from mitochondria to the nerve terminals in the brain. (De Graaf, 2007). As tCr is not naturally produced in the brain, its concentration is assumed to be stable with no changes reported with age or a variety of diseases, and is used as a reference value for calculating metabolite ratios (e.g., NAA/Cr, tCho/Cr, etc.). Nevertheless, the use of tCr as a reference metabolite should be used with caution as decreased tCr levels have been observed in the chronic phases of many pathologies including tumors (Howe et al., 2003; Ishimaru et al., 2001), stroke (Gideon et al., 1992), and gliosis (van der Graaf, 2010).

5.3.4 Glutamate and Glutamine

Glutamate (Glu–*2.15 ppm*) and Glutamine (Gln–*2.45 ppm*) together form a complex of peaks (Glx complex) between 2.15 ppm and 2.45 ppm. Their distinction at 1.5T is difficult due to their similar chemical structures. However, at ≥3T Glu and Gln can be resolved (Srinivasan et al., 2006) and at magnetic fields of 7T and higher, the Glu and Gln resonances are visually separated, leading to considerable quantification accuracy (De Graaf, 2007).

Glu is the major excitatory neurotransmitter in the mammalian brain and the direct precursor for the major inhibitory neurotransmitter, γ-aminobutyric acid (GABA). Besides these roles, Glu is also an important component in the synthesis of other small metabolites (e.g., Glutathione) as well as larger peptides and proteins. The amino acid Gln, which is primarily located in astroglia, is involved in intermediary metabolism and is synthesized from Glu (De Graaf, 2007). The Glx complex plays a role in detoxification and regulation of neurotransmitters. Increased levels of Glx complex are markers of epileptogenic processes (Hammen et al., 2003; Simister et al., 2009) and decreased levels of Glx have been observed in Alzheimer, dementia and patients with chronic schizophrenia (Kantarci et al., 2003; Théberge et al., 2003). Glx complex increments have also been observed in the peritumoral brain edema correlated with neuronal loss and demyelination (Ricci et al., 2007). As reported by Srinivasan et al. (2005), Glx might be used as an in vivo index of inflammation since they observed elevated Glx levels in acute MS plaques but not in chronic ones.

5.3.5 N-Acetyl Aspartate

N-Acetyl Aspartate (NAA–*2.02 ppm*) in ^{1}H-MR spectra of normal brain, is the most prominent resonance originating from the methyl group of NAA at 2.02 ppm with a contribution from neurotransmitter N-aspartyl-glutamate (NAAG) (Frahm et al., 1991). NAA is exclusively localized in the central and peripheral nervous system synthesized in brain mitochondria. Its concentration varies subtly in different parts of the brain (Baker et al., 2008; Doelken et al., 2009; Pouwels and Frahm, 1998) but undergoes large developmental changes, almost doubling from birth to adulthood (Kreis et al., 1992; Toft et al., 1994; Barker, 2001). Although NAA is definitely considered a neuronal marker representing neuronal density and viability, its exact function remains largely unknown. Possible neuronal functions of NAA include osmoregulation (Taylor et al., 1995) and a breakdown product of NAAG (Blakely and Coyle, 1988; Martin et al., 2001).

The utility of NAA as an axonal marker is supported by the loss of NAA in many white matter diseases, including leukodystrophies (Morita et al., 2006; Távora et al., 2007), multiple sclerosis (MS) (Mader et al., 2008; Wattjes et al., 2008) and hypoxic encephalopathy (Rosen and Lenkinski, 2007), chronic stages of stoke (Mader et al., 2008; Saunders, 2000) and tumors (Howe et al., 2003; Law et al., 2002; Nelson et al., 2003; Sibtain et al., 2007; Soares and Law, 2009).

However, there are cases when the abnormal levels of NAA do not reflect changes in neuronal density, but rather a perturbation of the synthetic and degradation pathways of the NAA metabolism. For instance, in Canavan's disease, high levels of intracellular NAA (Barker et al., 1992; Lin et al., 2005) are due to deficiency of aspartoacylase (ASPA), which is the enzyme that degrades NAA to acetate and aspartate. Further examples that show the lack of direct relationship of NAA to neuronal integrity include various pathologies such as temporal lobe epilepsy (TLE) (Chernov et al., 2009; Hajek et al., 2008; Lee et al., 2005) or amyotrophic lateral sclerosis (ALS) (Sarchielli et al., 2001; Wang et al., 2006), which exhibit spontaneous or treatment reversals of NAA to normal levels (Barker et al., 2001).

NAA has been also detected in non-neuronal cells, such as oligodendrocytes, which may contribute to the overall NAA signal observed in a ^{1}H-MRS spectrum, suggesting that it may not be specific only for neuronal processes (Bhakoo and Pearce, 2000).

5.3.6 Lactate and Lipids

Lactate and free Lipids (Lac–*1.35 ppm*, Lip–*0.9–1.3 ppm*), should be maintained below or at the limit of detectability within the normal brain spectrum, overlapping with macromolecule (MM) resonances at 1.33 ppm (doublet) and 0.9–1.3 ppm, respectively. Any detectable increase of lactate and lipid resonances can therefore be considered abnormal.

Lac is present in both intracellular and extracellular spaces and can be considered an index of metabolic rate being the end-product of anaerobic glycolysis (Howe et al., 2003). Hence, increased lactate levels can been observed in a wide variety of conditions in which oxygen supply is restricted, such as acute and chronic ischemia (Graham et al., 1992; Mader et al., 2008; Saunders, 2000), metabolic disorders (Chi et al., 2011; Soares and Law, 2009; Cecil, 2006), and tumors (De Graaf, 2007; Howe et al., 2003; Nelson et al., 2003; Sibtain et al., 2007; Soares and Law, 2009).

Lactate also accumulates in tissues that have poor washout, like cysts (Chang et al., 1998; Mishra et al., 2004) and normal pressure hydrocephalus (Kizu et al., 2001). The aforementioned spectral region between 0.9 ppm and 1.3 ppm represents the methylene (1.3 ppm) and the methyl (0.9 ppm) groups of fatty acids. Fractured proteins and lipid layers become visible only during the breakdown of membranes. Regardless of the exact mechanism, an elevation of lipid resonances indicates cerebral tissue destruction such as infarction (Graham et al., 1992; Mader et al., 2008; Saunders, 2000), acute inflammation (Hayashi et al., 2003; Yeh et al., 2008), and necrosis (Ishimaru et al., 2001; Lai et al., 2008).

5.3.7 Less Commonly Detected Metabolites

The signals of most other metabolites in a brain spectrum are considerably smaller than those mentioned so far, often due to their splitting into multiplets and/or overlapping peaks. Some of these compounds are present in healthy brain tissue but because they are too small, their detection is difficult. Some examples of these compounds include Alanine, Glycine, Taurine, Glutathione, and several other amino acids such as Succinate, Acetate, Valine and Leucine.

Alanine (Ala–*1.47 ppm*) is an amino acid present in the normal brain, resonating at 1.47 ppm. It is frequently considered a specific metabolic characteristic of meningiomas, however, its identification rate varies from 32% to 100% (Bulakbasi et al., 2003; Chernov et al., 2011; Cho et al., 2003; Demir et al., 2006; Howe et al., 2003; Möller-Hartmann et al., 2002). It can also be present in neurocytomas (Krishnamoorthy et al., 2007), gliomas (Majós et al., 2004), and PNETs (Majós et al., 2002). In vivo ^{1}H-MRS at lower filed strengths often cannot provide a distinction between Ala and Lac peaks as they resonate in neighboring frequencies. When both metabolites are present, they produce a triplet peak located between 1.3 ppm and 1.5 ppm (Yue et al., 2008) observed at 3T and higher.

Glycine (Gly–*3.55 ppm*) is considered an inhibitory neurotransmitter (Yeh et al., 2008) and a possible antioxidant, distributed mainly in astrocytes and glycinergic neurons. It resonates at 3.55 ppm, overlapping with mI; therefore, its detection is impossible in a non-processed spectrum. Only in cases of mI absence, even low Gly levels can be quantified (Lehnhardt et al., 2005).

High levels of Gly have been observed in glioblastomas, medulloblastomas, ependymomas (Kinoshita and Yokota, 1997) and neurocytomas (Yeh et al., 2008). Some research groups have reported that this metabolite may provide a noticeable metabolic feature for the distinction of GBMs from lower grade astrocytomas (Kinoshita and Yokota, 1997; Lehnhardt et al., 2005), primary gliomas from recurrence (Lehnhardt et al., 2005) and glial tumors from metastatic brain tumors (Kinoshita and Yokota, 1997; Majós et al., 2003).

Taurine (Tau–*3.25 ppm and 3.42 ppm*) gives two triplets at 3.25 ppm and 3.42 ppm, significantly overlapping with Cho and mI. Tau is an inhibitory neurotransmitter that activates GABA-a receptors or strychnine-sensitive glycine receptors, and it has also been suggested as an osmoregulator and a neurotransmitter action modulator (Shirayama et al., 2010).

In a healthy brain, Tau is heterogeneously distributed throughout the brain, with the higher levels observed in the olfactory bulb, retina and cerebellum (Huxtable, 1989).

High levels of Tau have been observed in medulloblastoma (Panigrahy et al., 2006), pituitary adenoma and metastatic renal cell carcinoma (Kinoshita and Yokota, 1997). Increased levels of Tau have been also reported in the medial prefrontal cortex in schizophrenic patients (Shirayama et al., 2010).

Glutathione (GSH–*2.9 ppm*) is the major protective molecule of living cells resonating at 2.9 ppm. It has an important role against oxidative stress, serving as an antioxidant and detoxifier (An et al., 2009), and plays a role in apoptosis and amino acid transport (Opstad et al., 2003). This metabolite has been reported to participate in acute ischemic stroke patients as ischemia is associated with significant oxidative stress (An et al., 2009), and in other neurodegenerative diseases such as Parkinson's disease (Sian et al., 1994). GSH has also been found to be significantly elevated in meningiomas when compared to other tumors (Kudo et al., 1990; Opstad et al., 2003), showing, as well, an inverse relationship with glioma malignancy (Couldwell et al., 1992; Kudo et al., 1990).

Finally, several other amino Acids such as Succinate at 2.4 ppm, Acetate at 1.92 ppm, Valine and Leucine at 0.9 ppm, together with Alanine and Lactate, are the major spectral findings of bacterial and parasitic diseases. Acetate and Succinate presumably originate from enhanced glycolysis of the bacterial organism (Chang et al., 1998; Sibtain et al., 2007). The amino acids Valine and Leukine are known to be the end-products of proteolysis by enzymes released in pus (Sibtain et al., 2007). Specifically, Leucine and Valine peaks have been detected in cystercercosis lesions but not in spectra of brain tumors (Sibtain et al., 2007).

5.4 MRS Quantification and Data Analysis

Quantification:

- Metabolite ratios are considered robust and reproducible in a clinical environment, but are based on the assumption that total Cr remains constant, which might not be the case.
- Possible abnormal metabolite ratios may be the result of changes in both the numerator and denominator.
- Absolute quantification allows detection of metabolite abnormalities that might not be otherwise evaluated.
- Absolute quantification requires internal or external references, and correction for voxel contamination.
- Post processing:
  - Several specialized processing steps are necessary to improve the visual appearance of the MR spectrum or the accuracy during metabolites estimation.

5.4.1 Quantification

To produce a high-quality spectrum, the first step is to obtain the maximum signal strength of each metabolite in a given spectrum and determine its concentration. Of course, several specialized processing steps by sophisticated computer algorithms are performed on the acquired FID, either in the time domain or the frequency domain (i.e., before or after the Fourier transform). These processes are usually not seen by the user since they are performed by automated programs (different vendors use different programs), but play a major role on the quality of analyses. Quantitative analyses of the (spectral) data, as well as the methods for this analysis arguably have the same importance as the techniques used to collect them. Possible incorrect data analysis or potential artifacts and pitfalls may lead to systematic errors and hence misinterpretation of the clinical results.

The quantification of spectra is of paramount importance in in-vivo MRS since visual evaluation alone is obviously not enough. In terms of quantification, the "holy grail" of spectral analysis is to determine the concentrations of the compounds present in the human brain and evaluate them accordingly. That is, to determine the area under the spectral peak, which should be proportional to the metabolite concentration. Nevertheless, this process can be very challenging due to several reasons, such as resonance overlap, baseline distortions, and poorly approximated lineshapes with conventional models such as Gaussian or Lorentzian functions.

On the other hand, spectra can be simply quantified in terms of metabolite ratios rather than metabolite concentrations and it is the most common approach. This sounds safer since metabolite ratios are calculated by dividing the area of each metabolite peak by the area of a reference peak (commonly total Cr) from the same spectrum, thus providing a sound value. Consequently, this approach might remove systematic errors and provide reliable markers of tissue biochemistry. Unfortunately, the technique is based on the assumption that total Cr remains constant and this might not always be the case. Total Cr can be vastly modified across diseases or even during normal brain development or aging. Moreover, it is evident that possible abnormal metabolite ratios may be the result of changes in both the numerator and denominator.

Therefore, absolute metabolite concentrations can be considered better for in vivo spectra interpretation, despite the involved difficulties.

> So once again, inevitably the question arises: To quantify or not to quantify?

Unfortunately, there is no easy answer to this question. There are a variety of approaches and each one has its own set of advantages and disadvantages. The final decision ultimately depends on the clinical problem to be addressed, the experimental conditions used to obtain the data, and the analysis software available to the investigator. For a more comprehensive explanation of the quantification techniques and their pros and cons, please refer to the more detailed Chapter 6 (Section 6.2.8).

5.4.2 Post Processing Techniques

Again, in order to produce a high quality spectrum, several specialized processing steps can be performed retrospectively, improving the visual appearance of the MR spectrum or the accuracy during metabolite estimation. Therefore, for a reliable analysis of in vivo ^{1}H-MR spectra, an understanding of the principles of post-processing techniques is essential.

Signal post-processing can be performed either in the time domain or after Fourier transformation in the frequency domain (Tchofo and Baleriaux, 2009). The most common post-processing techniques for effective signal improvement are: (1) Eddy current correction, (2) removal of unwanted spectral components, (3) signal filtering, (4) zero filling, (5) phase correction, and (6) baseline correction.

Eddy current correction: During signal localization RF pulses are applied together with magnetic field gradients and additional crusher gradients are needed to eliminate the unwanted signal. The switching pattern of the gradients applied can cause eddy current (EC) artifacts that are time and space dependent, causing time dependent phase shifts in the FID and distorted metabolite lineshapes within the spectrum preventing accurate quantification (see also Chapter 2). In a spectrum, EC artifacts can be removed by acquiring an additional FID without water suppression. The phase of the water FID is determined in each time point and it is subtracted from the phase of the corrupted FID (van der Graaf, 2009). Alternative methods for EC correction without the need for an additional reference peak (e.g., water) have also been proposed making the phase-distorted signal in the spectrum resonate at a different frequency than the remaining components, or using the acquisition of two sequences with opposite gradients (Lin et al., 1994). The EC artifact correction comprises the first step of the post-processing procedure.

The removal of unwanted signals from the FID, which may disturb signals from the resonances of interest, is the next step of signal post-processing. The water signal is a typical example of such an unwanted peak. Water suppression during measurement is never perfect and a residual water signal remains in the spectrum, which often has a complicated lineshape (Jiru, 2008; van der Graaf, 2009). Residual water elimination from the FID can be achieved either by approximating the water signal and subtracting it from the FID (Pijnappel et al., 1992), by eliminating it using special filters (Coron et al., 2001; Sundin et al., 1999), or by applying baseline correction for the removal of the broad water peak from the spectrum (Jiru, 2008).

Quantitative analysis may be hampered by the existence of a distorted spectral baseline as the estimation of metabolite peak areas is not reliable. The baseline signal mostly comprises fast decaying components with very short T2* values such as macromolecules, the signal from the sample and residuals from inefficient water suppression. Therefore, baseline correction is crucial for robust data evaluation and quantification methods. Delayed acquisition (e.g., TE > 80 ms) can remove the macromolecules due to their shorter T2 relaxation times (~30 ms). But this has a significant drawback, namely the loss of information of many scalar-coupled resonances, which have been suggested to provide valuable information regarding tumor characterization (Castillo et al., 2000; Howe et al., 2003; Ricci et al., 2007; van der Graaf, 2010). Furthermore, macromolecule resonances can be reduced by utilizing the difference in T1 relaxation between metabolites and macromolecules (Behar et al., 1994). Additionally, the subtraction from the spectrum of the function, which describes the course of the distorted baseline, will efficiently cause the baseline to become straight.

Special functions, called filters, can be subsequently applied at the signal in the time domain. The goal is to enhance or suppress different parts of the FID, leading to improved signal quality. The three most commonly used filtering approaches are sensitivity enhancement, to reduce the noise from the FID; resolution enhancement, to achieve narrower metabolite linewidths; and apodization for a signal's ripple (due to signal truncation) reduction (Jiru, 2008).

The FID of a spectrum, when acquired, is sampled by the analog-to-digital converter over N points in accordance to the Nyquist sampling frequency. Therefore, if the number of points is not sufficient, the reliable representation of the signal fails. Instead of increasing the acquisition time with the inevitable noise increment, the acquired FID can artificially be extended by adding a string of points with zero amplitude to the FID prior to Fourier Transformation, a process known as zero filling. Zero filling does not increase the information content of the data but it can greatly improve the digital resolution of the spectrum. It also helps to improve the spectral appearance, rendering it an important post-processing step. After Fourier transformation, the spectrum will be phase corrected. When the zero-phased FID signal shifts to the frequency domain, it yields a complex spectrum with absorption (real) and dispersion (imaginary) Lorentz peaks. However, when the initial phase is non-zero, it is not attainable to restore pure absorption or dispersion line shapes and phase correction must be applied (Jiru, 2008). A zero-order phase correction compensates for any mismatch between the quadrature receive channels and the excitation channels to produce the pure absorption spectrum, whereas a first-order phase correction compensates for the nuclei dephase due to the delay between excitation and the detection of FID (Jiru, 2008).

After careful post-processing, the spectrum is presented to the user and it is ready for either a qualitative interpretation by calculating the metabolite ratios, or absolute quantification of the metabolites.

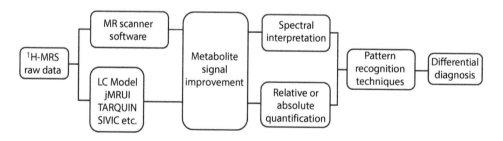

FIGURE 5.11 This diagram represents the complete MRS quantification and data analysis pipeline.

The aforementioned, post-processing techniques of the ^{1}H-MRS data can be performed either by programs available on commercial MR systems or by stand-alone software such as LC Model, jMRUI or others (Provencher, 2001; Preul et al., 1998; Naressi et al., 2001). In Figure 5.11 the diagram represents the complete MRS quantification and data analysis pipeline. For more details please refer to Section 6.2.9.

Consequently, the quantitative results can be used as an input in pattern recognition routines for classification purposes, leading to sounder interpretations and more efficient patient management (Tate et al., 2006; Tsolaki et al., 2014, 2015; Tsougos et al., 2012; Lukas et al., 2004).

5.5 Quality Assurance in MRS

Quality assessment in in-vivo MRS is of major importance for the establishment of MRS as a standard clinical method, ensuring reliability and reproducibility of a diagnosis. Similar to standard MRI examinations, quality assurance should be performed to ensure consistent high-quality data (Leach et al., 1995).

As described in the work of De Graaf (2007), quality assessment should consist of both MR system quality assurance (SQA) and quality control (QC) of spectral data acquired from patients and healthy volunteers. The system performance of the MR spectrometers must be checked bimonthly by a short measurement protocol using a specially designed phantom (e.g., INTERPRET or the spherical three-dimensional volume-elective test object, STO1, developed in a previous multicenter project on quality assessment in in-vivo MRS (Julià-Sapé et al., 2006). In addition, it is proposed that a more extended SQA protocol must be performed yearly and after each hardware or software upgrade.

The QC procedure of the MR spectra should comprise automatic determination of the signal-to-noise ratio (SNR) in a water-suppressed spectrum and of the line width of the water resonance (water band width, WBW) in the corresponding non-suppressed spectrum. Values of SNR > 10 and WBW < 8 Hz at 1.5T were determined empirically as conservative threshold levels required for spectra to be of acceptable quality. Moreover, a final QC check consisting of visual inspection of each clinically validated water-suppressed metabolite spectrum should be performed in order to detect artifacts such as large baseline distortions, exceptionally broadened metabolite peaks, insufficient removal of the water line, large phase errors, and signals originating from outside the voxel. These will be discussed in much more detail in the next chapter.

5.6 Conclusion

^{1}H-MRS can provide important in vivo metabolic information and significantly improve the overall diagnostic accuracy of the brain by complementing morphological findings from conventional MRI. Especially in the last 10 years, with the increasing availability of high field MR scanners (≥3T) this technique has proved to be an extremely valuable tool in solving difficult differential diagnostic problems, leading to more efficient patient management.

The way of further improvement should include the combination of ^{1}H-MRS with other advanced magnetic resonance techniques such as DWI/DTI and PWI, toward a complete multiparametric neuroimaging evaluation.

References

Alvarez-Linera, J. (2008). 3T MRI: Advances in brain imaging. *European Journal of Radiology, 67*(3), 415–426.

An, L., Zhang, Y., Thomasson, D. M., Latour, L. L., Baker, E. H., Shen, J., and Warach, S. (2009). Measurement of glutathione in normal volunteers and stroke patients at 3T using J-difference spectroscopy with minimized subtraction errors. *Journal of Magnetic Resonance Imaging, 30*(2), 263–270. doi:10.1002/jmri.21832

Avdievich, N., Pan, J., Baehring, J., Spencer, D., and Hetherington, H. (2009). Short echo spectroscopic imaging of the human brain at 7T using transceiver arrays. *Magnetic Resonance in Medicine, 62*(1), 17–25. doi:10.1002/mrm.21970

Baker, E. H., Basso, G., Barker, P. B., Smith, M. A., Bonekamp, D., and Horská, A. (2008). Regional apparent metabolite concentrations in young adult brain measured by1H MR spectroscopy at 3 Tesla. *Journal of Magnetic Resonance Imaging, 27*(3), 489–499. doi:10.1002/jmri.21285

Barker, P. B. (2001). N-acetyl aspartate--a neuronal marker? *Annals of Neurology, 49*(4), 423–424.

Barker, P. B., Boska, M. D., and Hearshen, D. O. (2001). Single-voxel proton MRS of the human brain at 1.5T and 3.0T. *Magnetic Resonance in Medicine, 45*(5), 765–769.

Barker, P. B., Bryan, R. N., Kumar, A. J., and Naidu, S. (1992). Proton NMR spectroscopy of Canavan's disease. *Neuropediatrics, 23*(5), 263–267.

Bartha, R., Drost, D., Menon, R., and Williamson, P. (2000). Comparison of the quantification precision of human short echo time 1H spectroscopy at 1.5 and 4.0 Tesla. *Magnetic Resonance in Medicine, 44*(2), 185–192. doi:10.1002/1522-2594(200008)44:2<185::aid-mrm4>3.3.co;2-m

Behar, K. L., Petroff, O. A., Rothman, D. L., and Spencer, D. D. (1994). Analysis of macromolecule resonances in 1H NMR spectra of human brain. *Magnetic Resonance in Medicine, 32*(3), 294–302.

Bhakoo, K. K. and Pearce, D. (2000). In vitro expression of N-acetyl aspartate by oligodendrocytes. *Journal of Neurochemistry, 74*(1), 254–262. doi:10.1046/j.1471-4159.2000.0740254.x

Blakely, R. D. and Coyle, J. T. (1988). The neurobiology of N-acetylaspartylglutamate. *International Review of Neurobiology, 30*, 39–100.

Boer, V. O., Siero, J. C., Hoogduin, H., Gorp, J. S., Luijten, P. R., and Klomp, D. W. (2011). High-field MRS of the human brain at short TE and TR. *NMR in Biomedicine, 24*(9), 1081–1088. doi:10.1002/nbm.1660

Bottomley P. A. and Park C. (1984). Selective volume method for performing localized NMR spectroscopy. U.S. patent, 4 480 228.

Brand, A., Richter-Landsberg, C., and Leibfritz, D. (1993). Multinuclear NMR studies on the energy metabolism of glial and neuronal cells. *Developmental Neuroscience, 15*(3–5), 289–298. doi:10.1159/000111347

Brown, T. R., Kincaid, B. M., and Ugurbil, K. (1982). NMR chemical shift imaging in three dimensions. *Proceedings of the National Academy of Sciences, 79*(11), 3523–3526. doi:10.1073/pnas.79.11.3523

Bulakbasi, N., Kocaoglu, M., Ors, F., Tayfun, C., and Uçöz, T. (2003). Combination of single-voxel proton MR spectroscopy and apparent diffusion coefficient calculation in the evaluation of common brain tumors. *AJNR. American Journal of Neuroradiology, 24*(2), 225–233.

Calli, C., Kitis, O., Ozel, A. A., Savas, R., Sener, R. N., and Yünten, N. (2002). Proton MR spectroscopy in the diagnosis and differentiation of encephalitis from other mimicking lesions. Journal of neuroradiology. *Journal de Neuroradiologie, 29*(1), 23–28.

Castillo, M., Kwock, L., and Smith, J. K. (2000). Correlation of myo-inositol levels and grading of cerebral astrocytomas. *AJNR. American Journal of Neuroradiology, 21*(9), 1645–1649.

Cecil, K. M. (2006). MR spectroscopy of metabolic disorders. *Neuroimaging Clinics of North America, 16*(1), 87–116, viii.

Chang, K. H., Han, M. H., Han, M. C., Jung, H. W., Kim, S. H., Kim, H. D., Song, I. C., and Seong, S. O. (1998). In vivo single-voxel proton MR spectroscopy in intracranial cystic masses. *AJNR. American Journal of Neuroradiology, 19*(3), 401–405.

Chernov, M. F., Kasuya, H., Nakaya, K., Kato, K., Ono, Y., Yoshida, S. et al. (2011). 1H-MRS of intracranial meningiomas: What it can add to known clinical and MRI predictors of the histopathological and biological characteristics of the tumor? *Clinical Neurology and Neurosurgery, 113*(3), 202–212. doi:10.1016/j.clineuro.2010.11.008

Chernov, M. F., Ochiai, T., Ono, Y., Muragaki, Y., Yamane, F., Taira, T. et al. (2009). Role of proton magnetic resonance spectroscopy in preoperative evaluation of patients with mesial temporal lobe epilepsy. *Journal of the Neurological Sciences, 285*(1–2), 212–219. doi:10.1016/j.jns.2009.07.004

Chi, C., Lee, H., Tsai, C., Chen, W., Tung, J., and Hung, H. (2011). Lactate peak on brain MRS in children with syndromic mitochondrial diseases. *Journal of the Chinese Medical Association, 74*(7), 305–309. doi:10.1016/j.jcma.2011.05.006

Cho, Y., Choi, G., Lee, S., and Kim, J. (2003). 1H-MRS metabolic patterns for distinguishing between meningiomas and other brain tumors. *Magnetic Resonance Imaging, 21*(6), 663–672. doi:10.1016/s0730-725x(03)00097-3

Christiansen, P., Henriksen, O., Stubgaard, M., Gideon, P., and Larsson, H. (1993). In vivo quantification of brain metabolites by 1H-MRS using water as an internal standard. *Magnetic Resonance Imaging, 11*(1), 107–118. doi:10.1016/0730-725x(93)90418-d

Coron, A., Vanhamme, L., Antoine, J., Hecke, P. V., and Huffel, S. V. (2001). The filtering approach to solvent peak suppression in MRS: A critical review. *Journal of Magnetic Resonance, 152*(1), 26–40. doi:10.1006/jmre.2001.2385

Couldwell, W. T., Antel, J. P., and Yong, V. W. (1992). Protein kinase C activity correlates with the growth rate of malignant gliomas. *Neurosurgery, 31*(4), 717–724. doi:10.1227/00006123-199210000-00015

Demir, M. K., Iplikcioglu, A. C., Dincer, A., Arslan, M., and Sav, A. (2006). Single voxel proton MR spectroscopy findings of typical and atypical intracranial meningiomas. *European Journal of Radiology, 60*(1), 48–55. doi:10.1016/j.ejrad.2006.06.002

De Graaf, R. A. (2007). *In Vivo NMR Spectroscopy Principles and Techniques.* Chichester, UK: Wiley.

De Zwart, J. A., Ledden, P. J., Gelderen, P. V., Bodurka, J., Chu, R., and Duyn, J. H. (2004). Signal-to-noise ratio and parallel imaging performance of a 16-channel receive-only brain coil array at 3.0 Tesla. *Magnetic Resonance in Medicine, 51*(1), 22–26. doi:10.1002/mrm.10678

Di Costanzo, A. D., Bonavita, S., Giannatempo, G. M., Nemore, F., Piccirillo, M., Scarabino, T. et al. (2003). High-field proton MRS of human brain. *European Journal of Radiology, 48*(2), 146–153.

Di Costanzo, A. D., Lechner, S. M., Popolizio, T., Schirmer, T., Scarabino, T., Trojsi, F. and Tosetti, M. (2007). Proton MR spectroscopy of the brain at 3 T: An update. *European Radiology, 17*(7), 1651–1662.

Doelken, M. T., Doerfler, A., Engelhorn, T., Hammen, T., Kloska, S., Mennecke, A. et al. (2009). Multi-voxel magnetic resonance spectroscopy of cerebral metabolites in healthy adults at 3 Tesla. *Academic Radiology, 16*(12), 1493–1501.

Drost, D. J., Riddle, W. R., and Clarke, G. D. (2002). Proton magnetic resonance spectroscopy in the brain: Report of AAPM MR Task Group #9. *Medical Physics, 29*, 2177–2197.

Duyn, J. H. and Moonen, C. T. (1994). Fast proton spectroscopic imaging of the human brain using multiple spin-echoes. *Magnetic Resonance in Medicine, 30*, 409–414.

Duyn, J. H., Gillen, J., Sobering, G., van Zijl, P. C., and Moonen, C. T. (1993). Multislice proton MR spectroscopic imaging of the brain. *Radiology, 188*, 277–282.

Edelstein, W. A., Glover, G. H., Hardy, C. J., and Redington, R. W. (1986). The intrinsic signal-to-noise ratio in NMR imaging. *Magnetic Resonance in Medicine, 3*(4), 604–618. doi:10.1002/mrm.1910030413

Frahm, J., Bruhn, H., Gyngell, M. L., Hänicke, W., Michaelis, T., and Merboldt, K. D. (1991). On the N-acetyl methyl resonance in localized 1H NMR spectra of human brain in vivo. *NMR in Biomedicine, 4*(4), 201–204.

Frahm, J., Bruhn, H., Gyngell, M. L., Merboldt, K. D., Hänicke, W., and Sauter, R. (1989). Localized high-resolution proton NMR spectroscopy using stimulated echoes: Initial applications to human brain in vivo. *Magnetic Resonance in Medicine, 9*(1), 79–93. doi:10.1002/mrm.1910090110

Gideon, P., Henriksen, O., Sperling, B., Christiansen, P., Olsen, T. S., Jorgensen, H. S., and Arlien-Soborg, P. (1992). Early time course of N-acetylaspartate, creatine and phosphocreatine, and compounds containing choline in the brain after acute stroke. A proton magnetic resonance spectroscopy study. *Stroke*, *23*(11), 1566–1572. doi:10.1161/01.str.23.11.1566

Gillard, J. H., Waldman, A. D., and Barker, P. B. (2010). *Clinical MR Neuroimaging: Physiological and Functional Techniques*. New York, NY: Cambridge University Press.

Graaf, R. A. (2007). *In vivo NMR Spectroscopy Principles and Techniques*. Chichester, UK: Wiley.

Graham, G. D., Blamire, A. M., Howseman, A. M., Rothman, D. L., Fayad, P. B., Brass, L. M. and Prichard, J. W. (1992). Proton magnetic resonance spectroscopy of cerebral lactate and other metabolites in stroke patients. *Stroke*, *23*(3), 333–340. doi:10.1161/01.str.23.3.333

Gruber, S., Mlynárik, V., and Moser, E. (2003). High-resolution 3D proton spectroscopic imaging of the human brain at 3 T: SNR issues and application for anatomy-matched voxel sizes. *Magnetic Resonance in Medicine*, *49*(2), 299–306. doi:10.1002/mrm.10377

Gu, H., Feng, H., Zhan, W., Xu, S., Silbersweig, D. A., Stern, E., and Yang, Y. (2002). *Single-shot interleaved Z-shim EPI with optimized compensation for signal losses due to susceptibility-induced field inhomogeneity at 3 T. NeuroImage*, *17*(3), 1358–1364. doi:10.1006/nimg.2002.1274

Haase, A., Frahm, J., Hanicke, W., and Matthaei, D. (1985). 1H NMR chemical shift selective (CHESS) imaging. *Physics in Medicine and Biology*, *30*(4), 341–344. doi:10.1088/0031-9155/30/4/008

Hajek, M., Dezortova, M., and Krsek, P. (2008). (1)H MR spectroscopy in epilepsy. *European Journal of Radiology*, *67*(2), 258–267.

Hammen, T., Stefan, H., Eberhardt, K. E., W-Huk, B. H., and Tomandl, B. F. (2003). Clinical applications of 1H-MR spectroscopy in the evaluation of epilepsies—What do pathological spectra stand for with regard to current results and what answers do they give to common clinical questions concerning the treatment of epilepsies? *Acta Neurologica Scandinavica*, *108*(4), 223–238. doi:10.1034/j.1600-0404.2003.00152.x

Hayashi, T., Fujihara, K., Higano, S., Jokura, H., Kumabe, T., Shiga, Y. et al. (2003). Inflammatory demyelinating disease mimicking malignant glioma. *Journal of Nuclear Medicine : Official Publication, Society of Nuclear Medicine*, *44*(4), 565–569.

Hetherington, H. P., Chu, W., Gonen, O., and Pan, J. W. (2006). Robust fully automated shimming of the human brain for high-field ^{1}H spectroscopic imaging. *Magnetic Resonance in Medicine*, *56*(1), 26–33. doi:10.1002/mrm.20941

Hoult, D. I. and Richards, R. E. (1976). The signal-to-noise ratio of the nuclear magnetic resonance experiment. *Journal of Magnetic Resonance*, *24*, 71–85.

Howe, F. A., Barton, S. J., Bell, B. A., Cudlip, S. A., Doyle, V. L., Griffiths, J. R. et al. (2003). Metabolic profiles of human brain tumors using quantitative in vivo 1H magnetic resonance spectroscopy. *Magnetic Resonance in Medicine*, *49*(2), 223–232.

Huxtable, R. J. (1989). Taurine in the central nervous system and the mammalian actions of taurine. *Progress in Neurobiology*, *32*(6), 471–533. doi:10.1016/0301-0082(89)90019-1

Hwang, T. and Shaka, A. (1995). Water suppression that works. Excitation sculpting using arbitrary wave-forms and pulsed-field gradients. *Journal of Magnetic Resonance, Series A*, *112*(2), 275–279. doi:10.1006/jmra.1995.1047

Ishimaru, H., Morikawa, M., Iwanaga, S., Kaminogo, M., Ochi, M., and Hayashi, K. (2001). Differentiation between high-grade glioma and metastatic brain tumor using single-voxel proton MR spectroscopy. *European Radiology*, *11*(9), 1784–1791. doi:10.1007/s003300000814

Jansen, J. F., Backes, W. H., Nicolay, K., and Kooi, M. E. (2006). 1H MR spectroscopy of the brain: Absolute quantification of metabolites. *Radiology*, *240*(2), 318–332. doi:10.1148/radiol.2402050314

Jiru, F. (2008). Introduction to post-processing techniques. *European Journal of Radiology*, *67*(2), 202–217.

Julià-Sapé, M., Acosta, D., Mier, M., Arùs, C., and Watson, D. (2006). A multi-centre, web-accessible and quality control-checked database of in vivo MR spectra of brain tumour patients. *Magnetic Resonance Materials in Physics, Biology and Medicine*, *19*(1), 22–33. doi:10.1007/s10334-005-0023-x

Kantarci, K., Boeve, B. F., Edland, S. D., Ivnik, R. J., Jack, C. R., Knopman, D. S. et al. (2003). Proton MR spectroscopy in mild cognitive impairment and Alzheimer disease: Comparison of 1.5 and 3 T. *AJNR. American Journal of Neuroradiology, 24*(5), 843–849.

Kararizou, E., Likomanos, D., Gkiatas, K., Markou, I., Triantafyllou, N., and Kararizos, G. (2006). Magnetic resonance spectroscopy: A noninvasive diagnosis of gliomatosis cerebri. *Magnetic Resonance Imaging, 24*(2), 205–207. doi:10.1016/j.mri.2005.10.032

Kinoshita, Y. and Yokota, A. (1997). Absolute concentrations of metabolites in human brain tumors using in vitro proton magnetic resonance spectroscopy. *NMR in Biomedicine, 10*(1), 2–12. doi:10.1002/(sici)1099-1492(199701)10:1<2::aid-nbm442>3.0.co;2-n

Kizu, O., Nishimura, T., and Yamada, K. (2001). Proton chemical shift imaging in normal pressure hydrocephalus. *AJNR. American Journal of Neuroradiology, 22*(9), 1659–1664.

Kreis, R. (1997). Quantitative localized 1H MR spectroscopy for clinical use. *Progress in Nuclear Magnetic Resonance Spectroscopy, 31*(2–3), 155–195. doi:10.1016/s0079-6565(97)00014-9

Kreis, R., Ross, B. D., Farrow, N. A., and Ackerman, Z. (1992). Metabolic disorders of the brain in chronic hepatic encephalopathy detected with H-1 MR spectroscopy. *Radiology, 182*(1), 19–27. doi:10.1148/radiology.182.1.1345760

Krishnamoorthy, T., Radhakrishnan, V. V., Thomas, B., Jeyadevan, E. R., Menon, G., and Nair, S. (2007). Alanine peak in central neurocytomas on proton MR spectroscopy. *Neuroradiology, 49*(7), 551–554. doi:10.1007/s00234-007-0224-2

Kruse, B., Hanefeld, F., Christen, H. J, Bruhn, H., Michaelis, T., Hänicke, W., and Frahm, J. (1993). Alterations of brain metabolites in metachromatic leukodystrophy as detected by localized proton magnetic resonance spectroscopy in vivo. *Journal of Neurology, 241*(2), 68–74. doi:10.1007/bf00869766

Kudo, H., Kokunai, T., Mio, T., Matsumoto, S., Sumino, K., and Tamaki, N. (1990). Quantitative analysis of glutathione in human brain tumors. *Journal of neurosurgery, 72*(4), 610–615.

Lai, P., Weng, H., Chen, C., Hsu, S., Ding, S., Ko, C. et al. (2008). In vivo differentiation of aerobic brain abscesses and necrotic glioblastomas multiforme using proton MR spectroscopic imaging. *American Journal of Neuroradiology, 29*(8), 1511–1518. doi:10.3174/ajnr.a1130

Leach, M., Collins, D., Keevil, S., Rowland, I., Smith, M., Henriksen, O. et al. (1995). Quality assessment in in vivo NMR spectroscopy: III. Clinical test objects: Design, construction, and solutions. *Magnetic Resonance Imaging, 13*(1), 131–137. doi:10.1016/0730-725x(94)00089-1

Lee, S. K., Kim, D. W., Kim, K. K., Chung, C. K., Song, I. C., and Chang, K. H. (2005). Effect of seizure on hippocampus in mesial temporal lobe epilepsy and neocortical epilepsy: An MRS study. *Neuroradiology, 47*(12), 916–923. doi:10.1007/s00234-005-1447-8

Lehnhardt, F., Bock, C., Röhn, G., Ernestus, R., and Hoehn, M. (2005). Metabolic differences between primary and recurrent human brain tumors: A ^{1}H NMR spectroscopic investigation. *NMR in Biomedicine, 18*(6), 371–382. doi:10.1002/nbm.968

Lin, A., Ross, B. D., Harris, K., and Wong, W. (2005). Efficacy of proton magnetic resonance spectroscopy in neurological diagnosis and neurotherapeutic decision making. *NeuroRX, 2*(2), 197–214. doi:10.1602/neurorx.2.2.197

Lin, C., Wendt, R. E., Evans, H. J., Rowe, R. M., Hedrick, T. D., and Leblanc, A. D. (1994). Eddy current correction in volume-localized MR spectroscopy. *Journal of Magnetic Resonance Imaging, 4*(6), 823–827. doi:10.1002/jmri.1880040614

Lukas, L., Devos, A., Suykens, J., Vanhamme, L., Howe, F., Majós, C. et al. (2004). Brain tumor classification based on long echo proton MRS signals. *Artificial Intelligence in Medicine, 31*(1), 73–89. doi:10.1016/j.artmed.2004.01.001

Luyten, P. R., Groen, J. P., Vermeulen, J. W., and Hollander, J. A. (1989). Experimental approaches to image localized human ^{31}P NMR spectroscopy. *Magnetic Resonance in Medicine, 11*(1), 1–21. doi:10.1002/mrm.1910110102

Luyten, P. R., Marien, A. J., Heindel, W., Gerwen, P. H., Herholz, K., Hollander, J. A. et al. (1990). Metabolic imaging of patients with intracranial tumors: H-1 MR spectroscopic imaging and PET. *Radiology, 176*(3), 791–799. doi:10.1148/radiology.176.3.2389038

Mader, I., Rauer, S., Gall, P., and Klose, U. (2008). 1H MR spectroscopy of inflammation, infection and ischemia of the brain. *European Journal of Radiology, 67*(2), 250–257. doi:10.1016/j.ejrad.2008.02.033

Majós, C., Alonso, J., Aguilera, C., Acebes, J. J., Arús, C., Gili, J. et al. (2003). Proton magnetic resonance spectroscopy ((1)H MRS) of human brain tumours: Assessment of differences between tumour types and its applicability in brain tumour categorization. *European radiology, 13*(3), 582–591.

Majós, C., Alonso, J., Aguilera, C., Acebes, J. J., Arús, C., Gili, J. et al. (2004). Brain tumor classification by proton MR spectroscopy: Comparison of diagnostic accuracy at short and long TE. *AJNR. American Journal of Neuroradiology, 25*(10), 1696–1704.

Majós, C., Alonso, J., Aguilera, C., Serrallonga, M., Acebes, J. J., Arús, C., and Gili, J. (2002). Adult primitive neuroectodermal tumor: Proton MR spectroscopic findings with possible application for differential diagnosis. *Radiology, 225*(2), 556–566. doi:10.1148/radiol.2252011592

Malhotra, H., Jain, K., Agarwal, A., Singh, M., Yadav, S., Husain, M. et al. (2009). Characterization of tumefactive demyelinating lesions using MR imaging and in-vivo proton MR spectroscopy. *Multiple Sclerosis Journal, 15*(2), 193–203. doi:10.1177/1352458508097922

Martin, E., Capone, A., Schneider, J., Hennig, J., and Thiel, T. (2001). Absence of N-acetylaspartate in the human brain: Impact on neurospectroscopy? *Annals of Neurology, 49*(4), 518–521. doi:10.1002/ana.102.abs

Mascalchi, M., Brugnoli, R., Guerrini, L., Belli, G., Nistri, M., Politi, L. S. et al. (2002). Single-voxel long TE 1H-MR spectroscopy of the normal brainstem and cerebellum. *Journal of Magnetic Resonance Imaging, 16*(5), 532–537. doi:10.1002/jmri.10189

Mescher, M., Tannus, A., Johnson, M., and Garwood, M. (1996). Solvent suppression using selective echo dephasing. *Journal of Magnetic Resonance, Series A, 123*(2), 226–229. doi:10.1006/jmra.1996.0242

Mishra, A. M., Behari, S., Gupta, R. K., Husain, N., Husain, M., Jaggi, R. S. et al. (2004). Role of diffusion-weighted imaging and in vivo proton magnetic resonance spectroscopy in the differential diagnosis of ring-enhancing intracranial cystic mass lesions. *Journal of Computer Assisted Tomography, 28*(4), 540–547.

Miyasaka, N., Takahashi, K., and Hetherington, H. P. (2006). Fully automated shim mapping method for spectroscopic imaging of the mouse brain at 9.4 T. *Magnetic Resonance in Medicine, 55*(1), 198–202. doi:10.1002/mrm.20731

Möller-Hartmann, W., Herminghaus, S., Krings, T., Lanfermann, H., Marquardt, G., Pilatus, U., and Zanella, F. E. (2002). Clinical application of proton magnetic resonance spectroscopy in the diagnosis of intracranial mass lesions. *Neuroradiology, 44*(5), 371–381.

Moonen, C. T., Kienlin, M. V., Zijl, P. C., Cohen, J., Gillen, J., Daly, P., and Wolf, G. (1989). Comparison of single-shot localization methods (steam and press) for in vivo proton NMR spectroscopy. *NMR in Biomedicine, 2*(5–6), 201–208. doi:10.1002/nbm.1940020506

Morita, H., Imamura, A., Matsuo, N., Tatebayashi, K., Omoya, K., Takahashi, Y., and Tsujino, S. (2006). MR imaging and 1H-MR spectroscopy of a case of van der Knaap disease. *Brain and Development, 28*(7), 466–469. doi:10.1016/j.braindev.2005.12.006

Naressi, A., Couturier, C., Castang, I., Beer, R. D., and Graveron-Demilly, D. (2001). Java-based graphical user interface for MRUI, a software package for quantitation of in vivo/medical magnetic resonance spectroscopy signals. *Computers in Biology and Medicine, 31*(4), 269–286. doi:10.1016/s0010-4825(01)00006-3

Nelson, S. J. (2003). Multivoxel magnetic resonance spectroscopy of brain tumors. *Molecular Cancer Therapeutics, 2*(5), 497–507.

Nelson, S. J., Vigneron, D. B., and Dillon, W. P. (1999). Serial evaluation of patients with brain tumors using volume MRI and 3D ^{1}H MRSI. *NMR in Biomedicine, 12*(3), 123–138. doi:10.1002/(sici)1099-1492(199905)12:3<123::aid-nbm541>3.0.co;2-y

Ocali, O. and Atalar, E. (1998). Ultimate intrinsic signal-to-noise ratio in MRI. *Magnetic Resonance in Medicine, 39,* 462–473. doi:10.1002/mrm.1910390317

Opstad, K., Provencher, S., Bell, B., Griffiths, J., and Howe, F. (2003). Detection of elevated glutathione in meningiomas by quantitative in vivo ^{1}H MRS. *Magnetic Resonance in Medicine, 49*(4), 632–637. doi:10.1002/mrm.10416

Panigrahy, A., Blüml, S., Finlay, J. L., Gonzalez-Gomez, I., Gilles, F. H., Krieger, M. D. et al. (2006). Quantitative short echo time 1H-MR spectroscopy of untreated pediatric brain tumors: Preoperative diagnosis and characterization. *AJNR. American Journal of Neuroradiology, 27*(3), 560–572.

Patt, S. L. and Sykes, B. D. (1972). Water eliminated Fourier transform NMR spectroscopy. *The Journal of Chemical Physics, 56*(6), 3182–3184. doi:10.1063/1.1677669

Pijnappel, W., Boogaart, A. V., Beer, R. D., and Ormondt, D. V. (1992). SVD-based quantification of magnetic resonance signals. *Journal of Magnetic Resonance (1969), 97*(1), 122–134. doi:10.1016/0022-2364(92)90241-x

Piotto, M., Saudek, V., and Sklenář, V. (1992). Gradient-tailored excitation for single-quantum NMR spectroscopy of aqueous solutions. *Journal of Biomolecular NMR, 2*(6), 661–665. doi:10.1007/bf02192855

Pouwels, P. J. and Frahm, J. (1998). Regional metabolite concentrations in human brain as determined by quantitative localized proton MRS. *Magnetic Resonance in Medicine, 39*(1), 53–60. doi:10.1002/mrm.1910390110

Preul, M. C., Caramanos, Z., Leblanc, R., Villemure, J. G., and Arnold, D. L. (1998). Using pattern analysis of in vivo proton MRSI data to improve the diagnosis and surgical management of patients with brain tumors. *NMR in Biomedicine, 11*(4–5), 192–200. doi:10.1002/(sici)1099-1492(199806/08)11:4/5<192::aid-nbm535>3.0.co;2-3

Provencher, S. W. (2001). Automatic quantitation of localized in vivo 1H spectra with LCModel. *NMR in Biomedicine, 14*(4), 260–264.

Ricci, R., Agati, R., Bacci, A., Battaglia, S., Leonardi, M., Maffei, M., and Tugnoli, V. (2007). Metabolic findings on 3T 1H-MR spectroscopy in peritumoral brain edema. *AJNR. American Journal of Neuroradiology, 28*(7), 1287–1291.

Rosen, Y. and Lenkinski, R. E. (2007). Recent advances in magnetic resonance neurospectroscopy. *Neurotherapeutics: The Journal of the American Society for Experimental NeuroTherapeutics, 4*(3), 330–345.

Ross, B. D. (1991). Biochemical considerations in ¹H spectroscopy. Glutamate and glutamine; Myo-inositol and related metabolites. *NMR in Biomedicine, 4*(2), 59–63. doi:10.1002/nbm.1940040205

Sarchielli, P., Alberti, A., Chiarini, P., Gobbi, G., Gallai, V., Presciutti, O. et al. (1999). Absolute quantification of brain metabolites by proton magnetic resonance spectroscopy in normal-appearing white matter of multiple sclerosis patients. *Brain: A Journal of Neurology, 122 (Pt 3)*, 513–521.

Sarchielli, P., Chiarini, P., Gobbi, G., Gallai, V., Pelliccioli, G. P., Presciutti, O., and Tarducci, R. (2001). Magnetic resonance imaging and 1H-magnetic resonance spectroscopy in amyotrophic lateral sclerosis. *Neuroradiology, 43*(3), 189–197.

Saunders, D. E. (2000). MR spectroscopy in stroke. *British Medical Bulletin, 56*(2), 334–345.

Shirayama, Y., Obata, T., Matsuzawa, D., Nonaka, H., Kanazawa, Y., Yoshitome, E., and Iyo, M. (2010). Specific metabolites in the medial prefrontal cortex are associated with the neurocognitive deficits in schizophrenia: A preliminary study. *NeuroImage, 49*(3), 2783–2790. doi:10.1016/j.neuroimage.2009.10.031

Sian, J., Agid, Y., Dexter, D. T., Daniel, S., Javoy-Agid, F., Jenner, P. et al. (1994). Alterations in glutathione levels in Parkinson's disease and other neurodegenerative disorders affecting basal ganglia. *Annals of neurology, 36*(3), 348–355.

Sibtain, N., Howe, F., and Saunders, D. (2007). The clinical value of proton magnetic resonance spectroscopy in adult brain tumours. *Clinical Radiology, 62*(2), 109–119. doi:10.1016/j.crad.2006.09.012

Simister, R. J., Mclean, M. A., Barker, G. J., and Duncan, J. S. (2009). Proton MR spectroscopy of metabolite concentrations in temporal lobe epilepsy and effect of temporal lobe resection. *Epilepsy Research, 83*(2–3), 168–176. doi:10.1016/j.eplepsyres.2008.11.006

Soares, D. P. and Law, M. (2009). Magnetic resonance spectroscopy of the brain: Review of metabolites and clinical applications. *Clinical radiology, 64*(1), 12–21.

Spielman, D. M., Pauly, J. M., Macovski, A., Glover, G. H., and Enzmann, D. R. (1992). Lipid-suppressed single-and multisection proton spectroscopic imaging of the human brain. *Journal of Magnetic Resonance Imaging, 2*(3), 253–262. doi:10.1002/jmri.1880020302

Srinivasan, R. (2005). Evidence of elevated glutamate in multiple sclerosis using magnetic resonance spectroscopy at 3 T. *Brain, 128*(5), 1016–1025. doi:10.1093/brain/awh467

Srinivasan, R., Cunningham, C., Chen, A., Vigneron, D., Hurd, R., Nelson, S., and Pelletier, D. (2006). TE-averaged two-dimensional proton spectroscopic imaging of glutamate at 3 T. *NeuroImage, 30*(4), 1171–1178. doi:10.1016/j.neuroimage.2005.10.048

Star-Lack, J., Nelson, S. J., Kurhanewicz, J., Huang, L. R., and Vigneron, D. B. (1997). Improved water and lipid suppression for 3D PRESS CSI using rf band selective inversion with gradient dephasing (basing). *Magnetic Resonance in Medicine, 38*(2), 311–321. doi:10.1002/mrm.1910380222

Stephenson, M. C., Francis, S. T., Gunner, F., Greenhaff, P. L., Macdonald, I. A., Morris, P. G. et al. (2011). Applications of multi-nuclear magnetic resonance spectroscopy at 7T. *World Journal of Radiology, 3*(4), 105–113.

Sundin, T., Vanhamme, L., Hecke, P. V., Dologlou, I., and Huffel, S. V. (1999). Accurate quantification of 1H spectra: From finite impulse response filter design for solvent suppression to parameter estimation. *Journal of Magnetic Resonance, 139*(2), 189–204. doi:10.1006/jmre.1999.1782

Tate, A. R., Acosta, D. M., Arús, C., Bosson, J. L., Bell, B. A., Cabañas, M. E. et al. (2006). Development of a decision support system for diagnosis and grading of brain tumours using in vivo magnetic resonance single voxel spectra. *NMR in Biomedicine, 19*(4), 411–434.

Távora, D. G., Nakayama, M., Gama, R. L., Alvim, T. C., Portugal, D., and Comerlato, E. A. (2007). Leukoencephalopathy with brainstem and spinal cord involvement and high brain lactate: Report of three brazilian patients. *Arquivos de Neuro-Psiquiatria, 65*(2b), 506–511. doi:10.1590/s0004-282x2007000300028

Taylor, D. L., Clark, J. B., Davies, S. E., Doheny, M. H., Obrenovitch, T. P., Patsalos, P. N., and Symon, L. (1995). Investigation into the role of N-acetylaspartate in cerebral osmoregulation. *Journal of Neurochemistry, 65*(1), 275–281.

Tchofo, P. J. and Balériaux, D. (2009). Brain (1)H-MR spectroscopy in clinical neuroimaging at 3T. Journal of neuroradiology. *Journal de Nuroradiologie, 36*(1), 24–40.

Théberge, J., Al-Semaan, Y., Williamson, P. C., Menon, R. S., Neufeld, R. W., Rajakumar, N. et al. (2003). Glutamate and glutamine in the anterior cingulate and thalamus of medicated patients with chronic schizophrenia and healthy comparison subjects measured with 4.0-T proton MRS. *American Journal of Psychiatry, 160*(12), 2231–2233. doi:10.1176/appi.ajp.160.12.2231

Tkác, I., Andersen, P., Adriany, G., Gruetter, R., Merkle, H., and Ugurbil, K. (2001). In vivo 1H NMR spectroscopy of the human brain at 7 T. *Magnetic Resonance in Medicine, 46*(3), 451–456.

Toft, P. B., Leth, H., Lou, H. C., Pryds, O., and Henriksen, O. (1994). Metabolite concentrations in the developing brain estimated with proton MR spectroscopy. *Journal of Magnetic Resonance Imaging, 4*(5), 674–680. doi:10.1002/jmri.1880040510

Tsolaki, E., Svolos, P., Kousi, E., Kapsalaki, E., Fezoulidis, I., Fountas, K. et al. (2015). Fast spectroscopic multiple analysis (FASMA) for brain tumor classification: A clinical decision support system utilizing multi-parametric 3T MR data. *International Journal of Computer Assisted Radiology and Surgery, 10*(7), 1149–1166. doi:10.1007/s11548-014-1088-7. Epub 2014 Jul 15.

Tsolaki, E., Svolos, P., Kousi, E., et al. (2013). Automated differentiation of glioblastomas from intracranial metastases using 3T MR spectroscopic and perfusion data. *International Journal of Computer Assisted Radiology and Surgery, 8,* 751. https://doi.org/10.1007/s11548-012-0808-0

Tsougos, I., Svolos, P., Kousi, E., Fountas, K., Theodorou, K., Fezoulidis, I., and Kapsalaki, E. (2012). Differentiation of glioblastoma multiforme from metastatic brain tumor using proton magnetic resonance spectroscopy, diffusion and perfusion metrics at 3T. *Cancer Imaging, 26*(12), 423–436. doi:10.1102/1470-7330.2012.0038

van der Graaf, M. (2010). In vivo magnetic resonance spectroscopy: Basic methodology and clinical applications. European Biophysics Journal, *39*(4), 527–540. http://doi.org/10.1007/s00249-009-0517-y

Wang, S., Desiderio, L. M., Elman, L. B., Krejza, J., McCluskey, L. F., Melhem, E. R. et al. (2006). Amyotrophic lateral sclerosis: Diffusion-tensor and chemical shift MR imaging at 3.0 T. *Radiology, 239*(3), 831–838.

Wang, Z., Zhao, C., Yu, L., Zhou, W., and Li, K. (2009). Regional metabolic changes in the hippocampus and posterior cingulate area detected with 3-Tesla magnetic resonance spectroscopy in patients with mild cognitive impairment and Alzheimer disease. *Acta Radiologica, 50*(3), 312–319. doi:10.1080/02841850802709219

Wattjes, M. P. and Barkhof, F. (2009). High field MRI in the diagnosis of multiple sclerosis: High field–high yield? *Neuroradiology, 51*(5), 279–292. doi:10.1007/s00234-009-0512-0

Wattjes, M. P., Bogdanow, M., Harzheim, M., Lutterbey, G. G., Schild, H. H., and Träber, F. (2008). High field MR imaging and 1H-MR spectroscopy in clinically isolated syndromes suggestive of multiple sclerosis: Correlation between metabolic alterations and diagnostic MR imaging criteria. *Journal of Neurology, 255*(1), 56–63.

Yeh, I. B., Lim, C. C., Ng, W. H., Xu, M., Ye, J., and Yang, D. (2008). Central neurocytoma: Typical magnetic resonance spectroscopy findings and atypical ventricular dissemination. *Magnetic Resonance Imaging, 26*(1), 59–64.

Yue, Q., Isobe, T., Shibata, Y., Anno, I., Kawamura, H., Yamamoto, Y. et al. (2008). New observations concerning the interpretation of magnetic resonance spectroscopy of meningioma. *European Radiology, 18*(12), 2901–2911. doi:10.1007/s00330-008-1079-6

Zhang, Y., Li, S., and Shen, J. (2009). Automatic high-order shimming using parallel columns mapping (PACMAP). *Magnetic Resonance in Medicine, 62*(4), 1073–1079. doi:10.1002/mrm.22077

6

Artifacts and Pitfalls of MRS

6.1 Introduction

Focus Point

- Subject movement can be a significant source of artifact in MRS.
- MRS is particularly sensitive to field inhomogeneity.
- Good quality spectra require careful pre-scanning and shimming, as well as good fat and water suppression.
- Optimal post processing and correct interpretation are key aspects in clinical MRS.
- Filtering, phase-correction, and baseline correction improve MRS data.
- Effects of Gadolinium on spectra are negligible.

Magnetic resonance spectroscopy (MRS) is a challenging technique and should only be used when spectral data accuracy is guaranteed, so as to provide reliable clinical information that cannot be obtained by other imaging techniques. If used correctly and accurately, it can be an invaluable tool for differential diagnosis and grading of tumors as well as for the evaluation of other neurodegenerative diseases.

Unfortunately, the use of MRS in clinical practice has been quite limited and this is due to several reasons. First, it is a time-consuming technique and requires a constant balance between time investment and quality of spectra. The total scan time is greatly affected by the preparation steps such as planning and pre-scanning for optimal water suppression, as well as the limitations regarding voxel size in the length of image acquisition (see also Chapter 5). Second, the technique is very sensitive to magnetic field inhomogeneities and frequently requires careful manual shimming to ensure field uniformity. Artifacts can arise from unexpected variables such as patient temperature variations, braces on teeth or hair gel, to more expected variations such as the proximity of the voxel to the sinuses and skull or normal regional variations. MRS is also very sensitive to motion; hence lengthy acquisitions are always a threat.

Another major difficulty for the clinical use of MRS is the lack of user's expertise (MRS specialists) and the fact that the entire procedure, from data acquisition and data processing to analysis and interpretation, is far from being an automated push-button technique. Although currently the data acquisition is largely automated, the processing and analysis is performed offline, usually based on non-commercial packages that are not integrated automatically into the scanners' workflow. Last but one of the more important limitations of MRS is that quality control is often nonexistent (Kreis, 2012). It follows that spectral analysis is arguably as important as data collection, and that incorrect analysis can lead to systematic errors and mis-interpretation of spectra leading to misdiagnosis. In fact, it seems that currently, there is a lack of widely accepted standards and spectral quality definitions for the acquisition and process-ing of MRS data, which could be applied to optimize the procedure.

To overcome these limitations, researchers and clinicians have tried to gain signal-to-noise ratio (SNR) and to detect additional metabolites more reliably by implementing higher field strengths, faster MRSI sequences, and motion corrected MRS acquisitions.

In spite of the aforementioned limitations, there are, of course, many clinical applications that have proved the considerable potential of MR spectroscopy in clinical practice, and this chapter is not intended to convey the idea that analysis and interpretation of MRS is not robust or unreliable. On the contrary the idea is to highlight potential pitfalls and artifacts both in the acquisition as well as in the analysis pipeline so that the reader can evaluate and avoid them, contributing to sounder interpretations.

6.2 Artifacts and Pitfalls

6.2.1 Effects of Patient Movement

Patient motion can be a very big problem, as in any other MRI sequence. Particularly for MRS, subject motion may have catastrophic effects, especially because these cannot be retrospec-tively corrected for. For example, considerable head movement may result in a volume of inter-est (VOI) shift, producing a spectrum that does not correspond to the initially planned volume. One way to verify a suspected VOI shift is to acquire a very fast axial acquisition after the completion of the spectrum and place it as a background image in the planned VOI to verify correct acquisition. If there is a considerable difference, spectroscopy planning should be cor-rected and repeated. The reason is that, besides wrong information with an erroneous voxel location, high magnetic susceptibility differences and contamination might be included in the VOI, such as tissue-air cavities, high water content (cysts), or hemorrhage and calcifications, resulting in poor shimming with lipid and water suppression problems, decreased SNR, and erroneous information. Therefore, patient movement should be strictly avoided if possible, or else (e.g., with children, patients who cannot cooperate or with high anxiety) sedation or anes-thesia should be used.

6.2.2 Field Homogeneity and Linewidth

The quality of a spectrum, that is, the ability to discriminate metabolite peaks or "resonances," absolutely depends on the linewidth of singlet peaks. The optimum (or the desired) peak line-width of a spectrum depends on the main magnetic field, B_0, and should be on the order of 0.05 parts per million (ppm). Parts per million is an arbitrary dimensionless unit that takes into account the different magnetic field strengths since locations in Hertz (Hz) would change with varying field strengths. Hence, the parts per million scale normalizes field strengths so that frequency can be calculated from the following equation:

$$\text{Frequency (Hz)} = \text{ppm} \times 42.6 \times \text{T} \tag{6.1}$$

For example, a metabolite with a singlet peak, such as creatine (including phosphocreatine Cr), corresponding to a methyl group that does not change chemical shift assignment and is used as a reference peak resonates at 3.0 ppm, which means 191.7 Hz at 1.5T and 383.4 Hz at 3T. In

Chapter 5 (Section 5.2.3), Figure 5.6 depicts the differences between 1.5T and 3.0T, emphasized by the overlapping of spectra. It is shown that at 3T, the SNR is about 25% higher and the spectral distance between the metabolites (in Hertz) is approximately doubled. The horizontal scale in an MR spectrum is usually displayed in ppm or in Hz while the vertical scale is usually displayed without values, automatically adjusted so that the largest peak fills the viewable area. Nevertheless, this might be misleading, especially when comparing different voxels within the same patient, corresponding to different scaling factors.

It follows that probably the single most important factor that affects the spectrum's quality is the main magnetic field's homogeneity. The measure of this quality is the linewidth of the spectral peaks, which should be less than ~5 Hz for 1.5T, or ~10 Hz for 3T. Linewidths broader than this may lead to serious degradation of spectral quality, leading to considerable misinterpretations (please refer to Figure 5.8 in Chapter 5). Several factors can cause broadening of the linewidths, including paramagnetic materials inside or close to the VOI, like shunts, post-surgical materials, hemorrhage and calcifications, which should be avoided. Moreover, areas of known susceptibility differences leading to poor field inhomogeneity should also be avoided if possible, including the mesial anterior temporal and inferior frontal lobes because of their proximity to air-cavities. This is why MRS of mesial temporal epilepsy is generally considered challenging. One may also encounter poor linewidths in certain anatomical areas of the brain like the caudate nucleus, globus pallidus and putamen, and substantia nigra because of the presence of heavy metals like iron.

Inadequate field homogeneity can only be overcome with appropriate shimming. This can be performed either manually or automatically. A shim failure (i.e., FWHM > 10 at 3T) can sometimes be improved by simply repeating the procedure, or by moving the VOI slightly and repeating the shimming procedure. Manual high order shimming requires a great deal of expertise and can be challenging; therefore, in a well maintained scanner, automatic shimming is considered a safer and more reproducible option. Nevertheless, there can be situations, when lesions are either too small or too heterogeneous (containing paramagnetic materials) or in very unfavorable locations, where MRS may simply not be safe to extract meaningful results.

Last, another factor that may influence the quality of spectra is magnetic field instability. It has been shown that the magnetic field (B_0) may encounter small drifts induced by gradient heating, which result in considerable resonance frequency shifts, causing poor linewidths due to suboptimal water suppression. This is a measurable effect, especially in high-field systems ($\geq 3.0T$) and can only be mitigated with dedicated correction algorithms (Ebel and Maudsley, 2005; Thiel et al., 2002).

6.2.3 Frequency Shifts and Temperature Variations

Another important parameter for the quality of MRS is to check the spectrum for frequency shifts, which may arise due to field inhomogeneities (see above) and, if necessary, to make manual corrections. Unfortunately, in most clinical systems, corrections are feasible only with the multivoxel option, or else post-processing software should be used (see Section 6.2.9). The reference peak used is usually the NAA peak at 2.02 ppm. It has to be noted that a frequency shift can also occur due to temperature variations. The MRS signal is inversely proportional to the absolute temperature of the tissue or test object (Tofts, 2003). The Boltzmann distribution gives a larger difference between spin populations as the temperature is reduced, and hence the magnetization of the sample increases as it becomes colder (Hoult and Richards, 2011). This effect must be taken into account, especially when using phantom measurements since the phantom temperature is at about 20° C (at scanner room temperature), which causes a shift in the spectrum of about 0.1 ppm to the right and needs to be corrected for in order to compare spectra. Figure 6.1 illustrates the temperature difference on a phantom measurement before and after correction.

6.2.4 Voxel Positioning

The choice of the position of the voxel is obviously of paramount importance for a meaningful in vivo ^{1}H-MRS, as it is important to locate the voxel in the appropriate region to detect the pathology of a lesion (Drost et al., 2002). For instance, there may be necrotic or cystic areas within a brain lesion that should be avoided. Moreover, active tumor and edema should be differentiated and the voxel should be accurately positioned, first on the most informative part of the lesion (the nodular part in the case of tumors), and second on the peritumoral area (Tsougos et al., 2012). Obviously, lesions do not always place themselves in convenient positions in the brain, nor in convenient shapes. But even in the case where a lesion has an absolutely symmetrical spherical shape, it is very important to realize the potential error that might occur from the incorrect voxel placement. Figure 6.2 illustrates a case of a hypothetical, completely

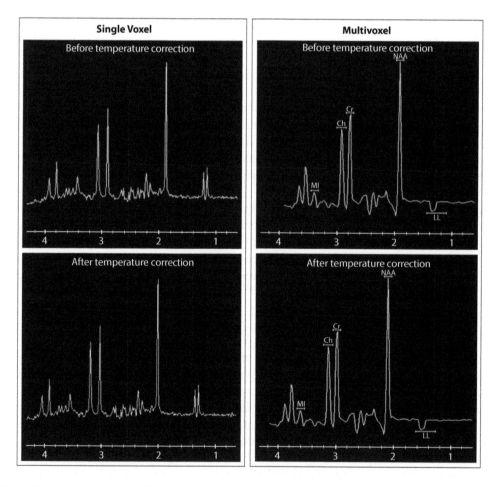

FIGURE 6.1 Single voxel (left) and Multivoxel spectroscopy (right) on a standard spectroscopy phantom (25-cm-diameter MRS HD sphere; General Electric Company) at 3.0T. The frequency shift before temperature correction (upper diagrams) and after temperature correction is noticeable.

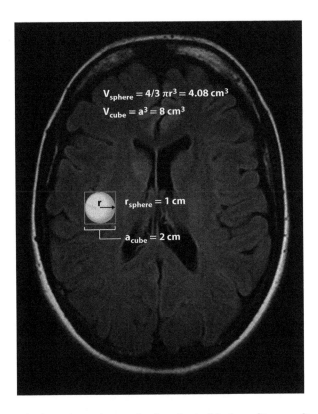

$$V_{sphere} = 4/3\,\pi r^3 = 4.08\ cm^3$$
$$V_{cube} = a^3 = 8\ cm^3$$

$$r_{sphere} = 1\ cm$$

$$a_{cube} = 2\ cm$$

FIGURE 6.2 A case of a hypothetical, completely spherical lesion of 1 cm radius. By placing a 2 × 2 × 2 cm³ voxel, it is evident that almost half of the voxel volume will cover peritumoral area leading to approximately 50% contamination from healthy or edematous tissue.

spherical lesion of 1 cm radius. By placing a voxel or cube of a 2-cm side on the center MR axial slice as seen in Figure 6.2, one would assume the coverage is more or less appropriate. Alas, taking into account the volume of the lesion (sphere) and the volume of the voxel (cube) it is evident that almost half of the voxel volume covers an area outside the lesion.

> Hence, the metabolic profile acquired would have approximately **50%** contamination from healthy or edematous tissue!

Moreover, it should be realized that there is a limitation regarding the minimum voxel size that can be used to obtain adequate SNR. Small voxels are easier to shim, but the signal depends on volume; so, a voxel of 1 cm³ (1 × 1 × 1 cm) is often considered the absolute minimum size to achieve a reasonable SNR for 3T, or 3.375 cm³ (1.5 × 1.5 × 1.5 cm) the practical minimum. If one wishes to further reduce the volume of the voxel, then the number of signal averages (or number of excitations [NEX]) recorded should be substantially increased.

This of course comes with a serious penalty: a substantial increase of scan time!

The SNR is approximately given by

$$SNR = SNR_1 NEX^{1/2} VOI \qquad (6.2)$$

where SNR_1 is the SNR for a single acquisition, NEX is the number of averages (excitations), and VOI is the volume of interest, from which the spectrum is being collected. From this equation, it is evident that SNR increases more with VOI and less with NEX. Therefore, to achieve a given SNR when VOI is reduced in half, NEX should be increased at least four times! Taking into account that the acquisition time can be estimated as NEX × TR, the total examination time should be increased four times, and this might not be possible during the clinical routine.

Correct voxel localization is also beneficial when monitoring brain pathology as well as treatment effects. As mentioned earlier, in cases of heterogeneous brain tumors, within a single lesion, there may be areas of necrosis, active tumor and edema. These areas are not always distinguishable on standard T1-weighted and T2-weighted MR images but rather post-contrast (Gd) images are required to ensure correct voxel placement in the active lesion. Therefore, spectroscopic prescription should always be used for adequate positioning. Thus, the possible effects of contrast on spectroscopy need to be anticipated (please refer to the next section).

6.2.5 Use of Contrast and Positioning in MRS

The controversy regarding the reliability of [1]H-MR spectroscopy findings in the presence of gadolinium chelates remains unanswered. It has been shown that gadolinium concentration has a minor effect in the signal of metabolites, causing a small amount of line broadening, so post-contrast T1 images can be useful for accurate voxel prescription (Lin and Ross, 2001; Murphy et al., 2002; Ricci et al., 2000; Taylor et al., 1995). On the other hand, this is true only if the Gd concentration of the lesion is relatively low: at higher concentrations, the T1 and T2 shortening effects should be taken into account. However, metabolite signals theoretically originate from the intracellular space where gadolinium cannot enter. Hence, the contrast agent effects reported in the literature (slight line broadening at typical, steady-state concentrations) have not been significant, although longer TEs seem to be more venerable than short TEs (Lin and Ross, 2001; Sijens et al., 1997; Sijens et al., 1998).

Figure 6.3 shows an example of MRS performed before and after Gd administration where post-contrast spectra are not significantly different relative to pre-contrast. The only necessary action is to readjust shimming post-Gd to account for any susceptibility differences.

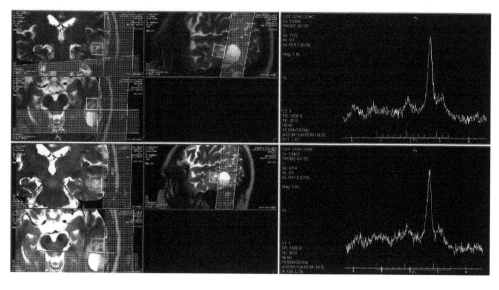

FIGURE 6.3   MRS performed before and after Gd administration. Post-contrast spectra are not significantly different relative to pre-contrast.

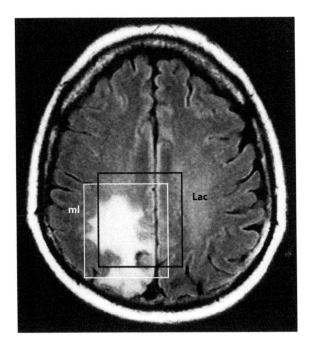

FIGURE 6.4 Chemical shift displacement (CSD) effect. Example of the calculated chemical shift displacement between mI and Lac at 3T. Considering all three directions, the mis-mapping can be as high as ~50%.

In conclusion, if magnetic field homogeneity can be adjusted, it appears unlikely that gadolinium can have any significant effect on brain spectra. The advent of more sophisticated medical imaging processing techniques has further improved voxel positioning within the lesion under study. Gajewicz et al. (2006) have shown that fused anatomical post-gadolinium MR imaging and single photon emission tomography (SPECT) imaging for ^{1}H-MRS, significantly improved the diagnosis of tumor grade and type. Moreover Dou et al. (2015) investigated the possibility for automatic voxel positioning for MRS at 7T, and concluded that, in view of the highly accurate and reproducible voxel alignment with automatic voxel positioning, the application of automatic rather than manual voxel positioning in future ultrahigh-field longitudinal MRS studies could be implemented.

Last, cautious spatial localization is used to remove unwanted signals from outside the ROI, like extracranial lipids and to avoid "partial volume effects," thereby providing a more genuine tissue characterization. Additional benefits from careful spatial voxel localization originate from the fact that variations in the main magnetic field and magnetic field gradients are greatly reduced, thereby providing narrower spectral lines and more uniform proton excitation.

6.2.6 Chemical Shift Displacement

The theoretical basis of MRS, the "chemical shift" phenomenon, can unfortunately be simultaneously the origin of an artifact called chemical shift displacement (CSD). This artifact is inherent to the use of magnetic field gradients for spatial localization via the spatial frequency dependency they create and cannot be completely avoided.

The nature of CSD artifacts between metabolites in MRS are identical to fat-water artifacts seen on conventional MRI. When a slice-selective pulse is applied, the position of the slice depends on the frequency offset and the gradient strength. The problem is that this frequency offset will be slightly different for each metabolite due to different resonance. It follows that the voxel position of each metabolite will also slightly vary, according to this difference (chemical

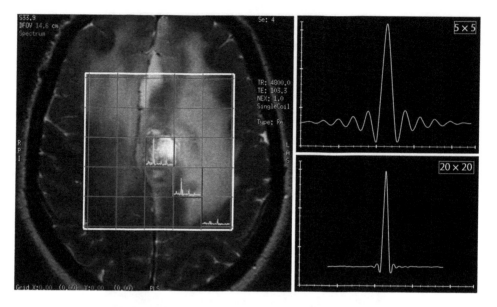

FIGURE 6.5 CSI multivoxel MRS technique. Voxel bleeding depends on the sampling and post-processing scheme of the PSF (right). By increasing the number of voxels (grid size) the PSF becomes sharper, resulting in reduced voxel bleeding.

shift) and the strength of the applied gradient. At the same time, the thickness of the slice will depend on the ratio of the RF pulse bandwidth to the gradient strength. This effect will lead to metabolite contributions that originate from partially differing spatial positions.

Consider, for example, a brain voxel containing lactate (Lac) and myo-inositol (mI) as seen in Figure 6.4. Their chemical shift difference is 2.2 ppm (3.5 ppm-1.3 ppm), which according to Equation 6.1, translates into a frequency difference of about 300 Hz at 3.0T. Considering a transmit RF-bandwidth at 1.5 kHz, then the two metabolites would be spatially dispersed or mis-mapped about 20% (300/1500) relative to one another. More importantly, it has to be stressed that because this displacement would occur in all three directions, only **51% ((1–0.20)^3)** overlap would occur in the entire voxel, even though the two resonances should coincide exactly (Elster, MRIquestions.com). This CSD artifact not only affects the relative positions of metabolites, but also their position relative to the prescribed planning grid and hence on the relevant anatomy. Chemical shift dispersion increases linearly with magnetic field strength, therefore artifacts are most problematic at higher magnetic fields.

As mentioned earlier, this effect unfortunately cannot be avoided completely. However, it can be minimized using the highest possible bandwidth RF pulses compared to the spectral dispersion (Kaiser et al., 2008; Scheenen et al., 2008). However, specific absorption rate (SAR) limitations are a major drawback since pulses tend to require high RF power levels, and may exceed maximum values for human applications. An alternative mitigating strategy is to use high bandwidth saturation pulses since these have higher bandwidths and lower SAR than either excitation or refocusing pulses (Edden and Barker, 2007; Edden et al., 2006; Tran et al., 2000).

6.2.7 Spectral Contamination or Voxel Bleeding

Contrary to single voxel spectroscopy, the main disadvantage of multi voxel spectroscopy or CSI is the presence of the so-called spectral contamination or voxel bleeding. This effect refers to the erroneous assignment of metabolite signals and therefore contamination of a voxel from adjacent locations. This artifact is due to the actual shape of the point spread function (PSF) associated with the limited spectroscopic grids (or matrix size) used, which are typically in the range of 8×8 to 32×32.

As seen in Figure 6.5, the shape of the PSF following Fourier transformation is such that the spectral peaks arising from a voxel can partially affect their neighboring voxels (spread out), although a significant truncation of the digitized signal must take place. The signal contribution (i.e., the voxel bleeding) depends on the sampling and post-processing scheme of the PSF. The exacerbation of voxel bleeding depending on low CSI grid size (matrix) is illustrated in the right part of Figure 6.5. The fewer the CSI voxels the broader the PSF, resulting in increased voxel bleeding.

This artifact cannot be completely eliminated, but it can be reduced by the use of digital filtering techniques in the spatial domain before Fourier transformation, but these result in an increase of the voxels' dimensions. This in turn can be compensated by increasing the grid, thereby increasing the number of voxels. By increasing the number of voxels, the PSF becomes sharper resulting in reduced voxel bleeding.

Another mitigating strategy to eliminate signals outside the VOI is the application of slice-selective pulses or outer volume saturation (OVS). These pulses are used to generate suppression slices on the periphery of the VOI so as to eliminate signals from extracranial water and lipids. Especially regarding lipid contamination, this is particularly important when the investigated pathology may involve Lactate signals (indicating necrosis). To positively identify Lac, a doublet positioned at 1.3 ppm should be accurately determined. Since Lac has a longer T2 compared to lipids, it may be better to use longer rather than shorter TEs since the Lac signal will appear inverted.

6.2.8 To Quantify or Not to Quantify?

Generally, there is a dispute in the MRS community regarding the issue of quantification versus visual interpretation and metabolite ratio evaluation. Obviously, the ultimate goal of MRS data analysis is to accurately determine metabolite concentrations based on the proportionality of metabolite resonance peak areas. Quantitative analyses of the (spectral) data, as well as the methods for this analysis arguably have the same importance as the techniques used to collect them. Possible incorrect data analysis or potential artifacts and pitfalls may lead to systematic errors and hence misinterpretation of the clinical results.

There can be different approaches regarding the quantification of spectra, and these can be broadly distinguished into the "relative" versus "absolute" quantification techniques.

6.2.8.1 Relative Quantification

The most widely used method of relative quantification is the use of an internal or endogenous metabolite, which can be used as the reference metabolite value. With this approach, each metabolite of interest in a given spectrum is normalized to the signal value of total Creatine (tCr - creatine plus phosphocreatine), which is commonly used as the reference metabolite since it is a strong signal with relatively low variability across brain regions and across subjects.

This technique is also called normalization (or "creatine normalization") because it reduces the variance due to the unknown scaling factor (see Section 6.2.2) in a given spectrum. That is, in any given acquisition, the different scaling factor would influence all metabolite signals equally; hence, any difference would cancel out when ratios are used. Obviously, these "metabolite ratios" do not reflect true concentrations unless the concentration of the reference metabolite is known.

Moreover, normalization removes another potential systematic error: the partial volume effect from the cerebrospinal fluid (CSF). Hence, although not absolute, normalization is still considered by many investigators the most robust, safe, and reproducible of all MRS quantitation techniques since it does not require extra time, software, or sequence modifications, and can be considered a reliable marker of tissue biochemistry. Nevertheless, the aforementioned normalization is only valid under the assumption that total creatine concentration does not vary across regions and subjects, which unfortunately is not the case. This variation across subjects is generally random, with a range of no more than 10%–15% (Webb et al., 1994) and is expected to decrease when many creatine-normalized metabolites are compared across different groups of patients.

Furthermore, relative quantification is associated with one more pitfall: the fact that an abnormal ratio might be equally the result of a change in the numerator, the denominator or even both, and therefore, ratios are intrinsically ambiguous and prone to systematic errors. For example, a reduced NAA/Cr ratio in the brain is interpreted as decreased NAA concentration, and therefore, associated with neuronal loss; however, a reduced NAA/Cr ratio may also be a result of increased Cr, which has been observed for instance in multiple sclerosis (Inglese et al., 2003).

6.2.8.2 Absolute Quantification

In order to resolve the uncertainties associated with the use of metabolite ratios, a variety of absolute quantification techniques have been devised.

6.2.8.2.1 Use of Water as an Internal Reference Signal

Since the water content in the brain tissue is almost constant, an alternative and widely used method for measuring metabolite concentrations is the use of unsuppressed tissue water signal as an internal reference (De Graaf, 2007; Ernst et al., 1993; Malucelli et al., 2009). Brain water content is well known and can be evaluated relatively easily by turning off the water suppression pulses (Christiansen et al., 1993; Thulborn and Ackerman, 1983). Moreover, comparing the water signal to creatine signal, the pathology-associated changes are relatively small. More specifically, the water content in gray and white matter varies about 10% (from 70% to 80%), and it is estimated that the proton concentration in the brain parenchyma is of the order of 77–88 M (M = molarity = mol/L). As in the case of creatine, the internal reference technique of water has the advantage of eliminating many potential systematic errors and pitfalls since it is acquired from the same VOI, with same sequence parameters, hence the metabolite concentration estimation becomes relatively accurate. Nevertheless, it must be noted that an additional structural MR image of the tissue within the voxel must be acquired to differentiate gray and white matter, and CSF, so that after coregistration, a segmentation of the voxel will follow (Gussew et al., 2012).

Obviously, there are several associated limitations and the use of water as an internal reference signal may be more complex than it appears. First, we have the basic assumption of the water content. This assumption is of course prone to error, particularly in unknown diseases and within focal lesions (Kreis, 2012). The differences in the water content of gray and white matter can be taken into account with segmentation information as already mentioned, but there is also the error due to unexpectedly high water contributions from CSF or cysts that may be included in the VOI due to the relatively large MRS voxels used (a practical minimum is of the order of 3.375 cm^3 at 3T). Also, brain water content may be variable in several cases, like in neonatal studies, or heterogeneous pathologies involving major changes in brain tissue.

6.2.8.2.2 Absolute Concentrations Using External Reference

Absolute concentrations using external reference involve the placement of a reference vial containing a known concentration, in a region outside the primary area of interest (e.g., next to the head) within the coil, simultaneously with the patient. The MR signal from this reference vial is measured, either before or after the brain spectrum is recorded, but, in any case, while the patient is still in the coil to ensure the same measurement condition, like loading and RF calibration. This way, the reference concentration is exactly known, allowing the measurement of the metabolite concentrations as well as the water content of the human brain with great precision. The disadvantage is that, because the reference vial is situated at a different spatial region compared to the MRS voxel, the reference signal intensity will be affected by inhomogeneities in the RF field. Another drawback of this technique is that the presence of the external vial may degrade the B0 field homogeneity within the brain due to magnetic susceptibility effects and is also susceptible to variations of the RF field. Last, some additional time is required for the second scan increasing the overall acquisition time.

6.2.8.2.3 Absolute Concentrations Using a Phantom

The use of a phantom or the "phantom replacement" technique is a variation of the external reference technique. It involves the acquisition of a spectrum from a phantom (comparable in size to a human head) of a reference metabolite of known concentration (e.g., creatine) before or after the patient has been imaged. In other words, the phantom replaces the head of the patient and the MRS scan is repeated with identical conditions. The advantage of this technique is that the voxel can be located at exactly the same position as in the patient (with respect to the coil) providing the same scaling factor, which can be calculated from the ratio of the metabolite peak integral with the known metabolite concentration in the phantom (Buonocore and Maddock, 2015). Obviously, various correction factors have to be applied; especially corrections taking into account the differences in RF coil loading that occur between the patient and the phantom since the main magnetic field should be re-shimmed to produce optimum spectra. One way of determining the loading factor is from the amount of RF power needed for a reference 90° pulse (Michaelis et al., 1993) or water suppression pulse (Danielsen and Henriksen, 1994). Finally, the extra time here is not an issue since it does not involve the patient. Summarizing all the aforementioned, Table 6.1 depicts the outline of all available MRS quantitation methods, listing their relative advantages and disadvantages.

TABLE 6.1   Summary of All Available MRS Quantitation Methods, Listing Their Relative Advantages and Disadvantages

Method	Advantages	Disadvantages
"Metabolite ratios"	Simple, no additional scan time, systematic errors, reproducible	Cannot quantify metabolite concentrations, pathological and regional variations of reference metabolite common
Spectrum from contralateral area	Normative data from subject, simple, no coil loading errors	Extra scan time required, sensitive to motion errors, not applicable in diffuse diseases, or midline structures
Internal water	Minimal extra scan time, simple, robust	Can be affected by pathological conditions
External reference	Accurate reference concentration	Extra scan time required, complicated due to susceptibility effects, B1 errors possible
Phantom replacement	No extra patient scan time required, accurate reference concentration, reduced susceptibility effects	RF coil loading correction required

6.2.9 Available Software Packages for Quantification and Analysis of MRS Data

A variety of software packages for the quantification and analysis of MRS data are available. In the following section, a brief informative description of the most commonly used packages is included so that the interested reader can proceed to a more comprehensive analysis, evaluation, and comparison. Most of these software packages are freely available, excluding the LCModel.

6.2.9.1 LCModel

One of the first software packages developed for MR automatic quantification of in vivo proton MR spectra and one of the most commonly used spectrum fitting tools is the so-called LCModel. It was established about 25 years ago (Provencher, 1993), and it was named after the analysis procedure it follows as a linear combination of model (LCModel) spectra, or basis spectra (Provencher, 2001). An example of an LCModel test run fit output is shown in Figure 6.6. The original data are shown in black, and the results of the LCModel fit in red with the metabolite ratios over Cr with their standard deviations on the right.

The advantage of the LCModel is that it is highly automated, requiring minimum user input compared to other software, especially in interdepartmental studies, where user dependency can be a major drawback for sound interpretations. Unfortunately, this is not freely available software and requires the purchase of a license, although there is a discount for nonprofit organizations. Nevertheless, it is the most widely used program, in part because of its commercial availability but also because phase, frequency and baseline correction is included in an almost fully automated procedure. Another important advantage of LCModel is the automatic

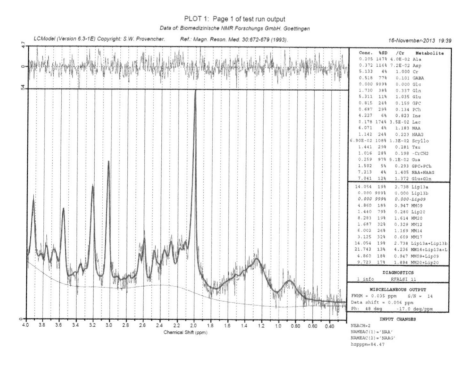

FIGURE 6.6 Example of the LCModel fit output. (Adapted from software's webpage: http://s-provencher.com/lcmodel.shtml.)

preprocessing of spectra, including built-in corrections for eddy current artifacts, which almost eliminates phase and fitting errors that might result from incorrect user parameter adjustments.

Generally, in order to improve fitting results, basis metabolite sets should be included in the system. These can be provided by the software vendor, although user generated basis sets, including macromolecules, is a safer and more robust choice. The fitting in LCModel is performed in the frequency and not in the time domain as in other software, with a data-based lineshape function, which theoretically takes into account different experimental conditions as it is nearly model-free. Another advantage of LCModel regarding CSI data analyses is that it first analyzes a central voxel of the subset and then works outwards, using Bayesian learning to get starting estimates and "soft constraints" for the first-order phase correction and the referencing shift from the preceding (often better) central voxels for the (often poorer) outer voxels. This speeds up and improves the overall analyses. LCModel is compatible with the majority of clinical MRS data formats.

6.2.9.2 jMRUI

jMRUI is a software package for advanced analysis of MRS and spectroscopic imaging (MRSI) data. The analysis is performed in the time-domain using signal processing algorithms to analyze and quantify spectroscopic data: from a single spectrum to a full multidimensional MRSI dataset (Naressi, 2001; Naressi et al., 2001; Stefan et al., 2009). The time domain is selected in order to limit data imperfections (signal truncation, echo timing errors, and spectral baseline) arising from FFT-processed data, with fewer assumptions and approximations.

The software interfaces algorithms for frequency selective filtering of signals, including both peak fitting and basis spectrum fitting (e.g., removal of residual water), correction of eddy-current artifacts, linear prediction, non-linear fitting (quantitation of metabolites), and many more.

The basic algorithm for basis spectra fitting is quantitation based on quantum estimation (QUEST) (Ratiney et al., 2004; Ratiney et al., 2005), and for peak fitting, one of the most commonly used peak fitting software packages for in vivo MRS, called AMARES (advanced method for accurate, robust, and efficient spectral fitting) (Vanhamme et al., 1997). QUEST fits the spectral data to a basis set in the time domain in a number of ways (e.g., truncation, FID weighting), while AMARES is based on a great wealth of prior knowledge constraints. Nevertheless, the entire spectral analysis can be performed using a set of mathematical model state-space fitting methods, using limited or even no prior knowledge. jMRUI is proprietary software distributed under its own license terms, and made freely available to registered users for non-commercial use. An example of the Cho/Cr metabolite map in a GBM-patient brain with jMRUI is illustrated in Figure 6.7.

Web Link: jMRUI, http://www.mrui.uab.es/mrui/mrui_Overview.shtml

6.2.9.3 TARQUIN

TARQUIN (totally automatic robust quantitation in NMR) is a freely distributed analysis tool for automatically determining the quantities of molecules present in in-vivo ^{1}H MRS and ex-vivo with ^{1}H HR-MAS spectroscopy data. The intended purpose of TARQUIN is to aid the characterization of pathologies, in particular brain tumors, based on a flexible time-domain fitting routine designed to give accurate, rapid and automated quantitation for routine analysis (Wilson et al., 2011). It includes a quantum mechanically based metabolite simulator to allow basis set construction optimized for the investigation of particular pathologies' sequence parameters. One unique feature included in TARQUIN is that it offers the possibility of

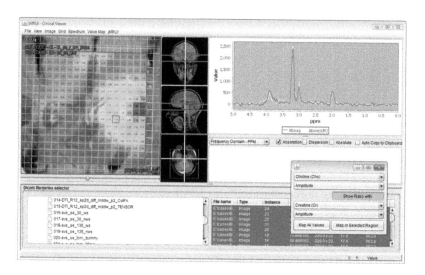

FIGURE 6.7   New jMRUI clinical viewer plug-in for MRS and MRSI: Cho/Cr metabolite map in a GBM-patient brain. (Adapted from software's webpage: http://www.mrui.uab.es/mrui/mrui_Overview.shtml.)

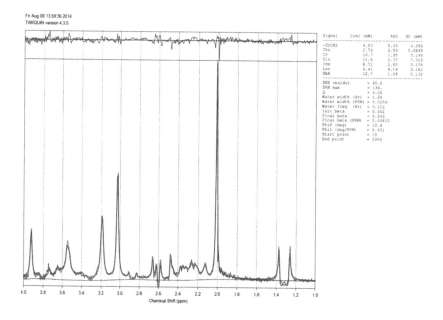

FIGURE 6.8   Example fit to the braino phantom (1.5T, PRESS TE = 30 ms) in TARQUIN. (Adapted from software's webpage: http://tarquin.sourceforge.net/.)

applying soft constrains on the metabolite peaks based on prior knowledge of relative concentrations for more accurate estimations. Moreover, it is compatible with most clinical MRS data in the standard DICOM format. An example fit to the braino phantom (1.5T, PRESS TE = 30 ms) is illustrated in Figure 6.8. Web Link: TARQUIN, http://tarquin.sourceforge.net/.

6.2.9.4 SIVIC

SIVIC is an open-source, standards-based software framework and application suite for processing and visualization of DICOM MRS data. It enables a complete scanner-to-PACS workflow for evaluation and interpretation of MRSI data. One unique feature of SIVIC is that it supports conversion of vendor-specific formats into the DICOM MR spectroscopy (MRS) standard, providing modular and extensible reconstruction and analysis pipelines, and tools to support visualization requirements associated with such data (Crane et al., 2013).

The benefit is that the entire exam including MRSI data and derived 3D metabolite maps can be archived in PACS. A key point here is that these 3D DICOM data are available for evaluation with any 3D imaging data in a multimodal evaluation scheme. An example is given in Figure 6.9, where the metabolite maps (bottom color overlay on a T1 contrast enhanced image) derived from MRSI data in SIVIC are exported as standard DICOM MR Image Storage SOP instances, which can be loaded into 3D DICOM image analysis software packages (shown here in 3D Slicer). The top panel shows ADC maps (color) on FLAIR images. SIVIC, http://sourceforge.net/projects/sivic/.

6.2.9.5 AQSES

AQSES (Automated Quantitation of Short Echo-time MRS Spectra) is a software package for quantitation of short echo time magnetic resonance spectra only, developed in JAVA (Poullet et al., 2007). The quantitation is approached in the time domain using the nonlinear least squares fitting, with a variable projection procedure. AQSES includes several preprocessing methods like phase-, frequency- and eddy current correction. Moreover, it includes a maximum-phase FIR filter or the HLSVD-PRO filter to remove unwanted components such as residual water. A macromolecular baseline is incorporated into the fit via nonparametric modeling, with penalized splines. This attribute can be viewed as a possible limitation since there is no prior knowledge of the macromolecule signals. Nevertheless, a graphical user interface for optimized quantitation is included and is therefore very suitable for spectroscopic imaging data. AQSES is freely available online from http://homes.esat.kuleuven.be/~biomed/software .php under an open source license.

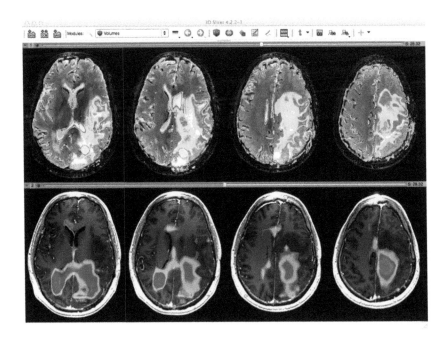

FIGURE 6.9 Example output from the SIVIC software. (Courtesy of Jason Crane, UCSF.)

6.3 Conclusion

Based on an extended review of the literature as well as personal clinical experience with 3T MR spectroscopy for the last 10 years, it is safe to argue that MRS can significantly improve the overall diagnostic accuracy of brain MRI, leading to safer and more robust clinical decisions, improving the efficiency of patient management. My personal opinion is that MRS should be integrated into a multi-parameter multi-modality approach of brain diseases for improved quantification accuracy.

Nevertheless, there should be an even greater degree of attention to the data analysis methods and new acquisition protocols used since the improved and advanced processing techniques also have the potential to optimize the clinical application of MRS. In spite of the deficiencies and potential artifacts and pitfalls of clinical MRS, it has proven to be a very powerful technique for the dynamic monitoring of brain tumor metabolism, and the progress in technology in the next few years is expected to increase the technique's application efficiency.

References

Buonocore, M. H. and Maddock, R. J. (2015). Magnetic resonance spectroscopy of the brain: A review of physical principles and technical methods. *Reviews in the Neurosciences, 26*(6), 609–632. doi:10.1515/revneuro-2015-0010

Christiansen, P., Henriksen, O., Stubgaard, M., Gideon, P., and Larsson, H. (1993). In vivo quantification of brain metabolites by 1H-MRS using water as an internal standard. *Magnetic Resonance Imaging, 11*(1), 107–118. doi:10.1016/0730-725x(93)90418-d

Crane, J. C., Olson, M. P., and Nelson, S. J. (2013). SIVIC: Open-source, standards-based software for DICOM MR spectroscopy workflows. *International Journal of Biomedical Imaging, 2013*, 1–12. doi:10.1155/2013/169526

Danielsen, E. R. and Henriksen, O. (1994). Absolute quantitative proton NMR spectroscopy based on the amplitude of the local water suppression pulse. Quantification of brain water and metabolites. *NMR in Biomedicine, 7*(7), 311–318. doi:10.1002/nbm.1940070704

De Graaf, R. A. (2007). *In Vivo NMR Spectroscopy Principles and Techniques.* Chichester, UK: Wiley.

Dou, W., Benner, T., Kaufmann, J., Li, M., Speck, O., Walter, M., and Zhong, K. (2015). Automatic voxel positioning for MRS at 7 T. *MAGMA, 28*(3), 259–270.

Drost, D. J., Clarke, G. D., and Riddle, W. R. (2002). Proton magnetic resonance spectroscopy in the brain: Report of AAPM MR Task Group #9. *Medical Physics, 29*(9), 2177–2197.

Ebel, A. and Maudsley, A. A. (2005). Detection and correction of frequency instabilities for volumetric ^{1}H echo-planar spectroscopic imaging. *Magnetic Resonance in Medicine, 53*(2), 465–469. doi:10.1002/mrm.20367

Edden, R. A. and Barker, P. B. (2007). Spatial effects in the detection of γ-aminobutyric acid: Improved sensitivity at high fields using inner volume saturation. *Magnetic Resonance in Medicine, 58*(6), 1276–1282. doi:10.1002/mrm.21383

Edden, R. A., Schär, M., Hillis, A. E., and Barker, P. B. (2006). Optimized detection of lactate at high fields using inner volume saturation. *Magnetic Resonance in Medicine, 56*(4), 912–917. doi:10.1002/mrm.21030

Elster, A. D. (2017). MRI questions and answers. MR Imaging Physics and Technology, http://www.mriquestions.com/.

Ernst, T., Kreis, R., and Ross, B. (1993). Absolute quantitation of water and metabolites in the human brain. I. Compartments and water. *Journal of Magnetic Resonance, Series B, 102*(1), 1–8. doi:10.1006/jmrb.1993.1055

Gajewicz, W., Grzelak, P., Gorska-Chrzastek, M., Zawirski, M., Kusmierek, J., and Stefanczyk, L. (2006). The usefulness of fused MRI and SPECT images for the voxel positioning in proton magnetic resonance spectroscopy and planning the biopsy of brain tumors: Presentation of the method. *Neurologia i Neurochirurgia Polska, 40*, 284–290.

Gussew, A., Erdtel, M., Hiepe, P., Rzanny, R., and Reichenbach, J. R. (2012). Absolute quantitation of brain metabolites with respect to heterogeneous tissue compositions in 1H-MR spectroscopic volumes. *Magnetic Resonance Materials in Physics, Biology and Medicine, 25*(5), 321–333. doi:10.1007/s10334-012-0305-z

Hoult, D. I. and Richards, R. E. (2011). The signal-to-noise ratio of the nuclear magnetic resonance experiment. 1976. *Journal of Magnetic Resonance, 213*(2), 329–343.

Inglese, M., Li, B. S., Rusinek, H., Babb, J. S., Grossman, R. I., and Gonen, O. (2003). Diffusely elevated cerebral choline and creatine in relapsing-remitting multiple sclerosis. *Magnetic Resonance in Medicine, 50*(1), 190–195. doi:10.1002/mrm.10481

Kaiser, L. G., Matson, G. B., and Young, K. (2008). Numerical simulations of localized high field 1H MR spectroscopy. *Journal of Magnetic Resonance, 195*(1), 67–75.

Kreis, R. (2012). Clinical MRS: Promise, Potential, Power and Pitfalls. *Proceedings of the International Society for Magnetic Resonance in Medicine, 20*.

Lin, A. P. and Ross, B. D. (2001). Short-echo time proton MR spectroscopy in the presence of gadolinium. *Journal of Computer Assisted Tomography, 25*(5), 705–712. doi:10.1097/00004728-200109000-00007

Malucelli, E., Manners, D. N., Testa, C., Tonon, C., Lodi, R., Barbiroli, B., and Iotti, S. (2009). Pitfalls and advantages of different strategies for the absolute quantification of N-acetyl aspartate, creatine and choline in white and grey matter by ^{1}H-MRS. *NMR in Biomedicine, 22*, 1003–1013. doi:10.1002/nbm.1402

Michaelis, T., Merboldt, K. D., Bruhn, H., Hänicke, W., and Frahm, J. (1993). Absolute concentrations of metabolites in the adult human brain in vivo: Quantification of localized proton MR spectra. *Radiology, 187*(1), 219–227. doi:10.1148/radiology.187.1.8451417

Murphy, P. S., Dzik-Jurasz, A. S., Leach, M. O., and Rowland, I. J. (2002). The effect of Gd-DTPA on T1-weighted choline signal in human brain tumours. *Magnetic Resonance Imaging, 20*(1), 127–130. doi:10.1016/s0730-725x(02)00485-x

Naressi, A. (2001). Java-based graphical user interface for the MRUI quantitation package. *Magnetic Resonance Materials in Biology, Physics, and Medicine, 12*(2–3), 141–152. doi:10.1016/s1352-8661(01)00111-9

Naressi, A., Couturier, C., Castang, I., Beer, R. D., and Graveron-Demilly, D. (2001). Java-based graphical user interface for MRUI, a software package for quantitation of in vivo/medical magnetic resonance spectroscopy signals. *Computers in Biology and Medicine, 31*(4), 269–286. doi:10.1016/s0010-4825(01)00006-3

Poullet, J., Sima, D. M., Simonetti, A. W., Neuter, B. D., Vanhamme, L., Lemmerling, P., and Huffel, S. V. (2007). An automated quantitation of short echo time MRS spectra in an open source software environment: AQSES. *NMR in Biomedicine, 20*(5), 493–504. doi:10.1002/nbm.1112

Provencher, S. W. (1993). Estimation of metabolite concentrations from localized in vivo proton NMR spectra. *Magnetic Resonance in Medicine, 30*(6), 672–679. doi:10.1002/mrm.1910300604

Provencher, S. W. (2001). Automatic quantitation of localized in vivo 1H spectra with LCModel. *NMR in Biomedicine, 14*(4), 260–264.

Ratiney, H., Coenradie, Y., Cavassila, S., Ormondt, D. V., and Graveron-Demilly, D. (2004). Time-domain quantitation of 1 H short echo-time signals: Background accommodation. *MAGMA Magnetic Resonance Materials in Physics, Biology and Medicine, 16*(6), 284–296. doi:10.1007/s10334-004-0037-9

Ratiney, H., Sdika, M., Coenradie, Y., Cavassila, S., Ormondt, D. V., and Graveron-Demilly, D. (2005). Time-domain semi-parametric estimation based on a metabolite basis set. *NMR in Biomedicine, 18*(1), 1–13. doi:10.1002/nbm.895

Ricci, P. E., Coons, S. W., Heiserman, J. E., Keller, P. J., and Pitt, A. (2000). Effect of voxel position on single-voxel MR spectroscopy findings. *AJNR. American Journal of Neuroradiology, 21*(2), 367–374.

Scheenen, T. W., Heerschap, A., Klomp, D. W., and Wijnen, J. P. (2008). Short echo time 1H-MRSI of the human brain at 3T with minimal chemical shift displacement errors using adiabatic refocusing pulses. *Magnetic Resonance in Medicine, 59*(1), 1–6.

Sijens, P. E., Bent, M. J., Dijk, P. V., Nowak, P. J., and Oudkerk, M. (1997). 1H chemical shift imaging reveals loss of brain tumor choline signal after administration of Gd-contrast. *Magnetic Resonance in Medicine, 37*(2), 222–225.

Sijens, P. E., Dijk, P. V., Levendag, P. C., Oudkerk, M. and Vecht, C. J. (1998). 1H MR spectroscopy monitoring of changes in choline peak area and line shape after Gd-contrast administration. *Magnetic Resonance Imaging, 16*(10), 1273–1280.

Taylor, J. S., Reddick, W. E., Kingsley, P. B., and Ogg, R. J. (1995). *Proton MRS after gadolinium contrast agent.* Third Scientific Meeting and Exhibition of the Society of Magnetic Resonance, In: Proceedings of the Society of Magnetic Resonance, Nice, France.

Thiel, T., Czisch, M., Elbel, G. K., and Hennig, J. (2002). Phase coherent averaging in magnetic resonance spectroscopy using interleaved navigator scans: Compensation of motion artifacts and magnetic field instabilities. *Magnetic Resonance in Medicine, 47*(6), 1077–1082. doi:10.1002/mrm.10174

Thulborn, K. R. and Ackerman, J. J. (1983). Absolute molar concentrations by NMR in inhomogeneous B1. A scheme for analysis of in vivo metabolites. *Journal of Magnetic Resonance (1969), 55*(3), 357–371. doi:10.1016/0022-2364(83)90118-x

Tofts, P. (2003). *Quantitative MRI of the Brain: Measuring Changes Caused by Disease.* Chichester, UK: John Wiley & Sons.

Tran, T. C., Vigneron, D. B., Sailasuta, N., Tropp, J., Roux, P. L., Kurhanewicz, J., and Hurd, R. (2000). Very selective suppression pulses for clinical MRSI studies of brain and prostate cancer. *Magnetic Resonance in Medicine, 43*(1), 23–33. doi:10.1002/(sici)1522-2594(200001)43:1<23::aid-mrm4>3.0.co;2-e

Vanhamme, L., Boogaart, A. V., and Huffel, S. V. (1997). Improved method for accurate and efficient quantification of MRS data with use of prior knowledge. *Journal of Magnetic Resonance, 129*(1), 35–43. doi:10.1006/jmre.1997.1244

Webb, P. G., Sailasuta, N., Kohler, S. J., Raidy, T., Moats, R. A., and Hurd, R. (1994). Automated single-voxel proton MRS: Technical development and multisite verification. *Magnetic Resonance in Medicine, 31*(4), 365–373. doi:10.1002/mrm.1910310404

Wilson, M., Reynolds, G., Kauppinen, R. A., Arvanitis, T. N., and Peet, A. C. (2011). A constrained least-squares approach to the automated quantitation of in vivo 1H magnetic resonance spectroscopy data. *Magnetic Resonance in Medicine, 65*(1), 1–12. doi:10.1002/mrm.22579

Functional Magnetic Resonance Imaging (fMRI)

7.1 Introduction

Focus Point

- Indirect measurement of neural activity, based on hemodynamic changes.
- Oxyhemoglobin is diamagnetic, while deoxyhemoglobin is paramagnetic.
- Following neuronal activity, there is an increase of oxygenated blood delivery, increasing oxyhemoglobin, therefore increasing the MR signal.
- Paradigm design for fMRI is challenging.
- Temporal resolution is limited by hemodynamic response.
- Need for cooperative patients.
- High magnetic fields (≥3 T) are preferable.
- Resting state fMRI is a recent concept.

7.1.1 What Is Functional Magnetic Resonance Imaging (fMRI) of the Brain?

Functional magnetic resonance imaging (fMRI) is a neuroimaging procedure performed in the MRI scanner to evaluate functional brain activity, basically by detecting changes associated with blood flow during specific stimuli. In that sense, the term "functional" may be considered misleading since the procedure actually provides an *indirect* measurement of neural activity, relying on the fact that cerebral blood flow and neuronal activation may be linked. Although fMRI is indeed one of the most recently applied methods of neuroimaging, the basic idea behind the technique is quite old. That is, if brain activity requires blood flow, it may be possible to estimate it by measuring changes in blood flow.

Interestingly, William James in *The Principles of Psychology*, a monumental text in the history of psychology published in 1890, mentioned an Italian scientist named Angelo Mosso who performed an experiment in the late 1800s by observing the patient on a delicately balanced table, which could tip downward either at the head or the foot if the weight of either end was increased. Theoretically, any emotional or intellectual activity of the subject would redistribute the blood flow and change the table's balance. Obviously, the results of this experiment were far from accurate but it proves that more than a century ago, Angelo Mosso was among the first to investigate the relationship between neural activity and cerebral hemodynamics (Mosso, 1881). Today, thanks to Seiji Ogawa, who sought to investigate the physiological condition of the brain in the late 1980s, fMRI uses the blood-oxygen-level dependent (BOLD) contrast, which is dependent on the content of deoxyhemoglobin in the blood. The idea was that the differing magnetic properties of deoxyhemoglobin and hemoglobin caused by blood flow to activated brain regions would cause measurable changes in the MRI signal (Ogawa and Lee, 1990; Ogawa et al., 1990).

More analytically, with the increase of neuronal activity, there is an increased demand for oxygen, which must be delivered to neurons by hemoglobin in capillary red blood cells. This increase results in an increase in blood flow to the regions of increased neural activity. The key is that hemoglobin is diamagnetic when oxygenated but paramagnetic when deoxygenated. Hence, the degree of oxygenation alters the local magnetic properties leading to small differences in the MR signal of the blood which can be used to detect brain activity.

fMRI is increasingly used in clinical practice, although it started mainly in the research world, where it was used to map brain activity evoked from certain stimuli or tasks (sensory, motor, cognitive, and emotional) in healthy individuals. More recently, this technique has evolved to study neurobehavioral disorders, such as Alzheimer's disease, epilepsy, traumatic brain injury, and brain tumors. Especially regarding tumors, the best developed clinical application of fMRI is pre-surgical mapping, which will be analytically discussed in a separate section at the end of this chapter. It has become popular since it is relatively easy to perform in the majority of existing MR scanners, it is reproducible, and does not involve the use of exogenous contrast agents. Moreover, relative to other functional imaging techniques (e.g., positron emission tomography) it is repeatable as it does not involve ionizing radiation, and yields superior temporal and spatial resolution.

7.1.2 Blood Oxygenation Level Dependent (BOLD) fMRI

Blood oxygenation level dependent (BOLD) imaging is the standard method used in functional MRI (fMRI) studies, and relies on the content of diamagnetic deoxyhemoglobin in the blood to delineate neural activity.

> But how is blood flow related to brain function?

Brain function requires a great deal of energy. Indicatively, the brain can reach 20% of the human body's oxygen consumption rate and 15% of its total blood flow. This energy is provided in the form of ATP, produced from glucose by oxidative phosphorylation. Hence, the rate of oxygen consumption can be assumed to be a good measure of the rate of energy consumption. The oxygen required by brain metabolism is supplied in the blood and therefore the high energy demand of the brain during certain tasks or stimuli results in increased oxygen delivery and increased blood flow, as illustrated in Figure 7.1.

Considering MR contrast, the important aspect is the presence of hemoglobin. Hemoglobin (Hb) is an iron-containing metalloprotein in the red blood cells used as an oxygen transport. From the contrast point of view, the hemoglobin molecule is an iron atom, bound in an organic structure. When no oxygen is bound to hemoglobin, it is called deoxyhemoglobin and it is paramagnetic (i.e., it has a small, positive susceptibility to magnetic fields), while when an oxygen molecule binds to hemoglobin, it becomes oxyhemoglobin and is diamagnetic (i.e., it has a weak, negative susceptibility to magnetic fields).

The paramagnetic properties of deoxyhemoglobin in blood vessels alter the local magnetic field causing a susceptibility difference between the vessel and its surrounding tissue. These susceptibility differences cause dephasing of the MR proton signal, reducing the T2* signal because the less uniform the field, the greater the number of different signal frequencies accumulated, and therefore the faster the signal decay. Thus, in a T2*-based sequence, the presence of deoxyhemoglobin in the blood vessels would cause the MRI signal to decay faster and therefore a darkening of the image, and vice versa. It follows that arterial blood (hence, an increase in diamagnetic oxyhemoglobin) produces more homogeneous magnetic field and therefore the MRI signal increases. More specifically, there is approximately a 0.08 ppm magnetic susceptibility difference between fully deoxygenated and fully oxygenated blood.

The overall effect is that when the brain is activated, tissue becomes more magnetically uniform with less susceptibility artifacts hence increased MR signal, as illustrated in Figure 7.1.

In fact, during activation (e.g., during a visual paradigm) deoxyhemoglobin is locally and temporarily increased since there is an increased demand for and consumption of oxygen by the cells, and therefore there should be a drop of the MR signal. Nevertheless, immediately after the stimulus, the local consumption of oxygen is compensated for by an increase in arterial blood flow leading to an increase in oxyhemoglobin. The change in blood flow is actually larger than that which is needed, hence, at the capillary level, there is an overall increase in oxygenated arterial blood versus deoxygenated venous blood (Ogawa and Lee, 1990; Ogawa et al., 1990).

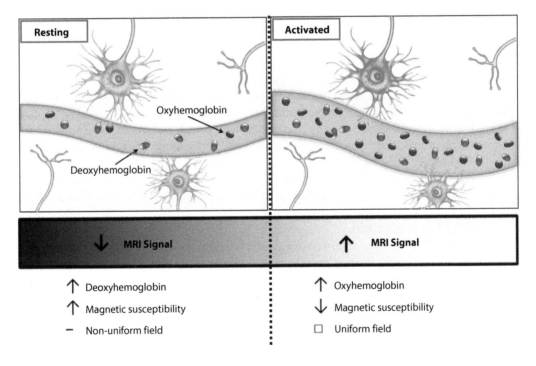

FIGURE 7.1 Basics of the BOLD effect in fMRI. When activated, (e.g., during certain stimuli) cells consume oxygen, thereby increasing deoxyhemoglobin's levels in the blood. Immediately after, an increase in arterial blood flow takes place to compensate for this consumption, leading to an increase in oxyhemoglobin. Since arterial blood is similar in its magnetic properties to tissue, the MR signals from activated regions will increase as magnetic susceptibility decreases.

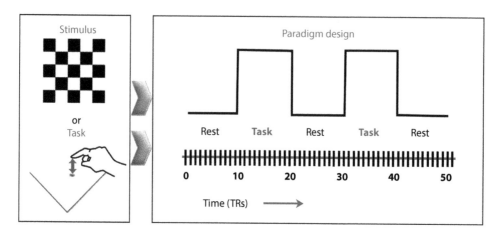

FIGURE 7.2   fMRI block design paradigm. A block design fMRI paradigm is designed so that a certain stimulus (e.g., flashing checkerboard) or task (e.g., finger tapping) in time "blocks" is interleaved with time blocks of rest, so that a considerable signal with repetition characteristics can be created and identified.

> Hence, oxygen alteration is the source of the MR signal since arterial blood (containing oxyhemoglobin), which is similar in its magnetic properties to tissue, restores the homogeneity of the magnetic field decreasing local magnetic susceptibility.

Why? Because when oxygen is bound to hemoglobin, the difference between the magnetic field applied by the scanner and that experienced by a blood protein molecule is less than when the oxygen is not bound. It should be noted that this signal increase is rather small (typically around 1% or maybe less). It follows that the term "activation" of the brain may be somewhat misleading since we are only talking about signal intensity changes of a few percent, and this explains why the data post-processing of fMRI experiments is difficult and still challenging. Obviously, the higher the magnetic field strength, the higher the BOLD signal recorded in fMRI, hence the easier the data post-processing. This is why fMRI is always preferable to be applied in high field scanners (≥3T).

Nevertheless, although the physical origins of BOLD signals are reasonably well understood, their precise connections to the underlying metabolic and electrophysiological activity need to be clarified further (Gore, 2003).

In any case, the fMRI procedure involves a great number of repetitions of scans, during which the subject carries out a specific task or is presented with a certain stimulus, overall called paradigms. Paradigms are usually arranged in a specific block design with altering periods of activity infused with periods of rest, or more rarely with periods of contrasting activity, so that a considerable signal with repetition characteristics can be created and identified. This is by far the most frequently used study design in clinical fMRI and is schematically illustrated in Figure 7.2.

It is evident that the success of the outcome depends on (1) the design and implementation of the paradigms, (2) the accuracy of the used MR scanning sequence, and (3) the accuracy and precision of data analysis.

7.1.3 fMRI Paradigm Design and Implementation

As mentioned earlier, neuronal activity is indirectly measured by evaluating the underlying physiological signals of the human brain. Obviously, this evaluation, and therefore the temporal resolution of fMRI, is going to be limited by the extent of methodological limitations compared to the physiological alterations aimed to be measured. In other words, the optimal choice of a good fMRI paradigm will be a compromise between the methodological advancements and technology capabilities of MRI measurements with the underlying pathophysiology.

> What we aim to measure is the so-called "hemodynamic response" or the relationship between the BOLD signal and neuronal activity.

Regarding efficient paradigm design, perhaps the most important limitation of the hemodynamic response is the delay of the signal following a certain neural activity, which can be of the order of seconds. Hence the response might exhibit considerable temporal blurring in relation to the underlying neuronal activity, thus bringing into question the reliability of the BOLD signal.

Indicatively, it was shown almost 20 years ago that even short periods of sensory stimulation (less than a second) that would be expected to result in proportionately short periods of neural activity, actually produce hemodynamic responses that can take place over a 10–12-s period and, in fact, have a delayed initiation of about 1 or 2 s (Boynton et al., 1996; Konishi et al., 1996). In fact, the underlying electrical activity (in milliseconds) is much quicker than the actual hemodynamic response (in seconds) on which the fMRI signal depends. Nevertheless, the information of the changes of neuronal activity can be extracted taking advantage of the temporal dynamics caused by certain block neuronal alterations, although the degree of temporal resolution is obviously rather poor.

Hence the smallest period of neural activity that can be reliably discriminated by fMRI (i.e., the temporal resolution) is of the order of 1 s, while the temporal resolution of EEG, for example, is of the order of milliseconds. There are several advanced techniques dedicated to the improvement of fMRI temporal resolution, mainly addressing the net magnetization recovery delay. For example, multiple coils can be used to speed up the acquisition time, or high pass filters can be used in k-space for selective data processing.

Coming back to efficient experimental design, the fundamental aim is to design a task that would determine as accurately as possible a specific hypothesis about a certain mental process. Many researchers call these mental processes brain functions of interest, or FOIs (Jezzard and Ramsey, 2004), where typically the FOIs are alternated with other tasks that do not trigger the process of interest, in a "on-off" design (Bandettini et al., 1993) since only two states are invoked. More recently a detailed quantitative characterization of the response strength of a certain region of interest (ROI) was exploited in order to characterize further how a known specialized brain area responds to subtle differences in carefully selected experimental conditions (ROI-based analysis) (Goebel, 2015).

7.1.3.1 Blocked versus Event-Related Paradigms

Another distinction between paradigms is the blocked versus the event-related. Historically, blocked designs were mainly adapted in fMRI from positron emission tomography (PET) studies. In a typical PET experiment, stimuli were clustered in certain "blocks" containing trials of the same event, and their mean activity was compared to another block. This was a necessity in PET due to the limited temporal resolution requiring blocks of about 1 min in length. The superior temporal resolution of fMRI allows block lengths of about 15 s (Maus et al., 2010).

Block designs, however, have some intrinsic fundamental limitations. Block designs cannot distinguish between correct and error trials (Taylor et al., 2007), and do not account for transient responses of the beginning and end of the task (Dosenbach et al., 2006; Fox et al., 2005). Finally, opposite responses (e.g., negative and positive) can be summed in a single block, thus averaging the overall response and degrading the response's magnitude (Meltzer et al., 2008).

On the other hand, in event-related paradigms the different condition trials are presented in random sequences and not grouped in blocks, provided sufficient time is given to separate the different responses. Event-related designs are advantageous compared to block designs, especially regarding the ability to avoid cognitive adaptation. The realization that underlying neuron activity can only be extracted from evoked hemodynamic responses and the need for improvement of the temporal resolution, urged researchers to use more complex task paradigms, that is, event-related designs. However, according to Petersen and Dubis (2012), it was quickly realized that this was not appropriate for hemodynamic responses from multiple trials that could overlap in time. Eventually, the most efficient event-related design was in fact a modified mixed block design.

7.1.3.2 Mixed Paradigm Designs

Before paradigm designs eventually transform to pulse sequences, they should fulfill some minimum criteria: (1) Obviously the first and more important is the high contrast-to-noise ratio (CNR), in the sense that they should be sensitive to potential mental changes; and (2) to give a true representation of the actual underlying procedure. More importantly, in order to give a true representation, the imaging experiment should interfere as little as possible with the paradigm, especially under the high levels of acoustic noise due to the time varying magnetic field gradients, which can go up to ~120 db for field strengths of ≥3T.

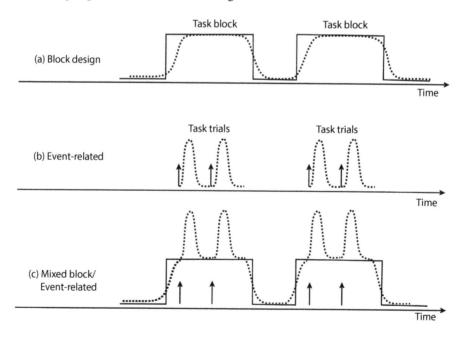

FIGURE 7.3 Different fMRI paradigm designs. (a) Block design yield a single magnitude reflecting all BOLD activity. (b) Event related design models separate trial types ignoring the sustained BOLD activity. (c) Mixed block/event-related design allows the simultaneous modeling of the task-related BOLD activity, as well as the transient, trial-related activity.

A fact that was not always taken into account until recently is that the sensitivity of the different types of paradigms may differ, depending on the signal changes that evolve with time. Axiomatically, the event-related designs are sensitive to transient changes in brain activity that are connected to the events of interest, while blocked designs are averaged over a certain period within the block, summing the event related changes in a single signal change. Nevertheless, the blocked paradigms may have additional sensitivity to less transient changes in activity that may persist for longer periods of time and that are not exclusively regulated by particular events.

In this regard, both paradigm designs, blocked and event-related, can be combined to investigate different situations, and their combination may prove to be beneficial to the overall fMRI experiment. Figure 7.3 demonstrates how the three different fMRI paradigm designs extract different signals from the hemodynamic response (BOLD activity).

It is evident that the mixed block/event-related design was developed to allow for simultaneous modeling of transient, trial-related and sustained task-related BOLD signals. The main advantage from the block design and separately event-related design is the possibility to evaluate BOLD signals related to task modes independent of the trials stimuli. Although very useful in some areas of investigation, like the memory and task control (Dennis et al., 2007; Dosenbach et al., 2006; Velanova et al., 2003), the usage of mixed paradigms has not become widespread, mainly because of the following limitations:

1. Poorly designed experiments may easily lead to misattribution of signals from different sources (Visscher et al., 2003).
2. The number of subjects evaluated is particularly important in this type of paradigm. Obviously determining how many subjects are needed for an experiment depends on the location and on the type of response that is modeled, but generally at least 25–30 subjects should be included (Petersen and Dubis, 2012).

In an attempt to summarize fMRI paradigm design and implementation, Table 7.1 presents all three designs with their relative advantages and disadvantages.

TABLE 7.1 All Three fMRI Paradigm Designs with Their Relative Advantages and Disadvantages

	Advantages	Disadvantages
Block design	-Robustness of results -Increased statistical power -Relatively large BOLD signal change related to baseline	-Induce differences in the cognitive "set" or strategies adopted by subjects -Difficult distinction between trial types within a block
Event-related design	-Analyses related to individual responses to trials -Less sensitivity to head motion artifacts -Randomization of the order of conditions presented -Variation of the time between stimulus presentations -Maintenance of a particular cognitive or attentional set	-Averaged responses -Decrease of signal-to-noise ratio
Mixed design	- Allows for extraction of transient and sustained BOLD activity - Different BOLD timescales suggest different neural functions	-Involves more assumptions than other designs -Power considerations

7.2 fMRI Acquisitions—MR Scanning Sequences

The goal of fMRI analysis is to examine changes in brain activity using the changes of the deoxyhemoglobin levels in brain tissue as described earlier. This goal, as any evaluation using MRI, is largely dependent on the scanning sequences and imaging parameters used. It should be clear by now that the phenomenon investigated is the so-called BOLD contrast, in which the presence or absence of deoxyhemoglobin induces changes in the T2* and T2 signal.

The scanning sequences should therefore be sensitive to changes in the chosen contrast and produce as high contrast-to-noise ratio (CNR) as possible. They should also give images that are a true representation of the actual underlying procedure. As mentioned earlier in this chapter, in order to give a true representation, the sequences should interfere as little as possible with the paradigm, especially under the high levels of acoustic noise due to the time varying magnetic field gradients, which can go up to ~120 db for field strengths of ≥3T.

7.2.1 Spatial Resolution

Spatial resolution of an fMRI study refers to the minimum discrimination ability between adjacent locations. As in any MRI study, it is measured by the size of a three-dimensional rectangular cuboid, the voxel. The voxel dimensions are determined by the slice thickness, and the area of the slice, as well as the matrix of the slice set by the scanning protocol. Excluding some specialized high-resolution studies, the voxel size will be typically in the 2–4 mm range dependent on the chosen contrast and the main magnetic field strength (Norris, 2015). The smaller the voxel, the fewer the neurons included, the less the blood flow, and hence, the lower the number of signals compared to larger voxels. Moreover, since scanning time is proportional to the number of voxels per slice and to the number of slices, smaller voxels are also more time consuming. In general, higher times in an MRI procedure lead to patient discomfort and loss of signal and should be avoided if possible. It is useful to realize that a 2–4-mm voxel would approximately contain a few million neurons and tens of billions of synapses, while the ideal neural activity signal would arise from the deoxyhemoglobin contribution to the BOLD phenomenon form the capillaries near the area of activity (Huettel et al., 2009). Obviously, a precise relationship between the voxel size and contrast cannot be simply calculated since it is largely dependent on the shimming outcome. Generally, matching the voxel volume to the cortical thickness—about 3 mm—can be considered a safe common practice (Bandettini et al., 1993).

7.2.2 Temporal Resolution

Temporal resolution of an fMRI study refers to the minimum period of neural activity reliably evaluated. The temporal resolution is largely dependent on the repetition or sampling time TR. The required temporal resolution must be tailored to the theoretical brain processing time for various events and pursued experiment. A TR in the range of 2–3 s is, at most static field strengths, almost ideal in terms of sensitivity as it is sufficient to allow near full recovery of the longitudinal magnetization between excitations, giving close to the maximum transverse magnetization at each excitation (Norris, 2015). Obviously, there is a broad range of temporal resolution demand as different processing procedures have various time spans. For example, the photoreceptors of the retina register a signal within a millisecond, while the neuronal activity related to the act of seeing lasts for more than 100 ms and emotional or physiological changes may last minutes.

These requirements for spatial and temporal resolution and the basic requirement for whole brain coverage elevate the necessity of high speed in data acquisition and set high temporal demands to about 10 slices per second. Moreover, another demanding aspect of fMRI protocol acquisition is that the scanner should be capable of acquiring and storing a rather huge number of multislice brain volumes as a time series (approx. 100–500), and that all these data should be reconstructed and analyzed. Until recently not many commercial scanners had this capability, and off-line reconstruction was used.

7.2.3 Pulse Sequences Used in fMRI

Nowadays it is widely accepted that the selection of the optimum sequence for fMRI experiments is a multi-variable and multi-level problem, with many different solutions, which require advanced knowledge and experience and can be demanding even for the experts of the field. Since the fMRI inception in early 1990s, there has been a rapid rate of improvement in every aspect of its workflow pipeline, including pulse sequence design. Moreover, the optimum sequence selection also largely depends on the contrast mechanism contribution and more importantly the static magnetic field in which it is going to be acquired.

Due to the mechanisms contributing to the BOLD signal, T2*—contrast sequences are theoretically expected to give the maximum sensitivity in an fMRI experiment. In the late 1970s, Sir Peter Mansfield introduced the fastest single-shot imaging sequence called Echo Planar Imaging, or EPI. Hence, Gradient-Echo EPI (GE-EPI) has been the most popular chosen sequence for fMRI, as depicted in Figure 7.4.

According to Menon et al. (1993), it is a straightforward algebraic exercise to show that under the assumption of exponential decay, the optimum echo time (TE) for fMRI is equal to T2*,

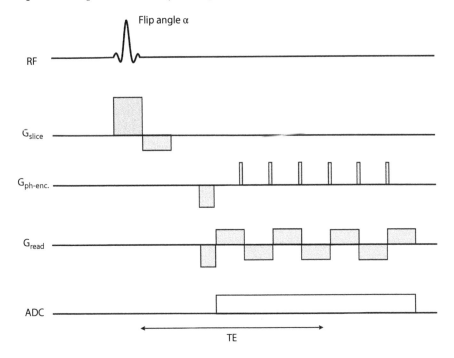

FIGURE 7.4 Multi-slice GE-EPI acquisition with the shortest possible inter-slice TR and a volume TR (that is the time until the same brain region is measured again) of less than 3 s.

which explains why it is necessary to incorporate an additional delay between the excitation and the signal acquisition. At 1.5T, T2* is about 50 to 60 ms, while at 3T, T2* is about 30 to 40 ms. As the static magnetic field increases, there is a proportional increase of the net magnetization and hence an increase of the signal-to-noise ratio (SNR). This increase of the SNR applies to both the anatomical image and to the magnitude of the activation-induced signal change. (Menon et al., 1993; Uğurbil et al., 1993).

The main disadvantage of the GE-EPI is the vulnerability to susceptibility effects from the gradients because of the necessarily long read-out gradient along the phase-encoding direction. Nevertheless, despite these difficulties (image distortions and signal voids), 2D EPI remains the most used sequence in fMRI, mainly by reducing the aforementioned artifact's effects by the application of multi-echo techniques.

According to Poser and Norris (2007), at higher field strengths, a good alternative to gradient-echo (GE) is the spin-echo (SE) sequence as the increased weighting toward the microvasculature results in intrinsically better localization of the BOLD signal. SE images are free of signal voids but the echo planar imaging (EPI) sampling scheme still causes geometric distortions. They conclude that multiply refocused SE sequences such as fast spin echo (FSE) are essentially artifact free but the major limitation is the high energy deposition, and long sampling times. For other investigators, the only feasible way to reduce EPI distortions is to reduce the duration of the readout using parallel imaging techniques (Griswold et al., 2002; Pruessmann et al., 1999), provided that multichannel receiver coils are used.

For the correction of distortions, a large number of methods have been developed, basically relying on the acquisition of a field map retrospectively applied to correct for distortions (Chen and Wyrwicz, 2001; Jezzard and Balaban, 1995; Zaitsev et al., 2004). On the other hand, a major limitation of such techniques is patient movement since it compromises the map's accuracy. For the interested reader, Weiskopf, Klose, Birbaumer, and Mathiak (2005) adequately analyzes image optimization using multi-echo EPI distortions and BOLD contrast for real-time fMRI.

Other popular approaches for fMRI acquisitions include the PRESTO (Principle of Echo Shifting with a Train of Observations) technique and the multi-echo approaches. PRESTO is basically an attempt to increase the efficiency of gradient echo based fMRI sequences using the delay required for the build-up of BOLD contrast (Liu et al., 1993; Liu et al., 1993).

Multi-echo approaches measure the signal at equal intervals during the free induction decay. In that sense, it is ensured that the measurement and analysis can be optimized for the signal decay characteristics on a voxel-wise basis. Moreover, one can have readout at short TEs where the signal would have disappeared, especially in regions with high susceptibility gradients.

Coming back to T2-weighted sequences, it has to be stated that they will always be less sensitive compared to T2* pulse sequences, especially for up to 3 T scanners. Nevertheless, as the magnetic field increases, and in particular at 7 T, which is increasingly used for fMRI experiments, T2-weighted fMRI seems to be the method of choice since it is better localized to the area of neuronal activity (Yacoub et al., 2005). The pulse sequences that can be used for T2-weighted fMRI are a simple modification of those previously described, based on EPI and fast spin echo.

More than a decade ago, turbo spin echo (TSE) pulse sequences have been suggested as an alternative to echo planar imaging (EPI) sequences for fMRI studies. Recently, Ye et al. (2010) reported the development of a modified half Fourier acquisition single-shot TSE sequence (mHASTE) with a three-fold GRAPPA, which improves temporal resolution, and the introduction of preparation time to enhance BOLD sensitivity. Using a classical flashing checkerboard block design, they systematically analyzed the BOLD signal characteristics of this novel method as a function of several sequence parameters and compared them with those of GE and SE EPI sequences. It was shown that mHASTE is more sensitive to extra-vascular BOLD effects around microvascular networks, thus enabling more accurate function localization. Compared to SE EPI, mHASTE has a ~50% reduction in activation cluster size and a ~20% decrease in BOLD contrast. However, a higher SNR and a spatially uniform temporal stability have been observed when using mHASTE compared with EPI sequences when scan times are held constant.

The mHASTE source image illustrates excellent image quality with no signal voids or image distortion. On the other hand, GE-EPI produced the most pronounced activation among the three sequences but some active voxels extended to areas outside the visual cortex, suggesting inaccuracy in the functional localization.

7.3 Analysis and Processing of fMRI Experiments

The aim of fMRI data analysis is to reliably correlate the detected hemodynamic response or brain activation with a specific task the subject is instructed to perform during the MRI scan. Furthermore, a major goal is to discover correlations with specific cognitive processes, such as memory and recognition, to specific areas and networks induced in the subject. The basic challenge is that the BOLD signal is relatively weak and additional sources of noise in the acquired data could further degrade it. Hence, after the fMRI experiment, the resulting data must be analyzed properly by performing a series of processing steps before the actual statistical analysis for the task-related brain activation can commence.

7.3.1 fMRI Datasets

In a typical fMRI experiment on a modern 3T scanner, the whole brain is scanned in a few seconds, acquiring a set (about 40) of "functional" images or "scans." These data are usually referred to as the *functional volume*. During an experiment, the subject usually performs certain tasks following a predefined protocol, and about 100 volumes are usually recorded. Each of these volumes consists of individual cuboids called voxels (see the previous section), including all the necessary information. Please note that the amount of raw data can go up to 500 MB per subject per session, which means up to several gigabytes per day of clinical routine.

As soon as the data of an individual subject is acquired, this data set should be prepared for statistical analysis since the aim of fMRI is to identify the areas of the brain where the signal detected is statistically significantly greater than noise.

The data analysis is currently performed in several software packages used for processing, analyzing, and displaying fMRI data, including the *FMRIB* software library from the Oxford group (http://www.fmrib.ox.ac.uk/fsl/), *Analysis of Functional NeuroImages* (http://afni.nimh. nih.gov/afni/) from the NIH, *Statistical Parametric Mapping* from the UCL group (http://www. fil.ion.ucl.ac.uk/spm/), BrainVoyager QX (http://www.brainvoyager.com), and others.

The general idea of fMRI signal processing is depicted in Figure 7.5. The process starts by convolving the predicted activity curve of the fMRI experiment, with a hemodynamic response function (HRF), producing the so-called predicted response. Each voxel contains a time-varying BOLD signal. Signals that match the predicted response (that is the modeled change in the BOLD signal) are identified as activation related to stimulus and can be processed for statistical analysis.

Nevertheless, the aforementioned general approach involves several other stages of preprocessing. The acquisition originally generates a 3D volume of the subject's brain at every scheduled repetition time (TR). Then, an array of voxel intensity values is generated, one per voxel in the scan. The next step is to unfold the 3D structure into a single line by arranging voxels

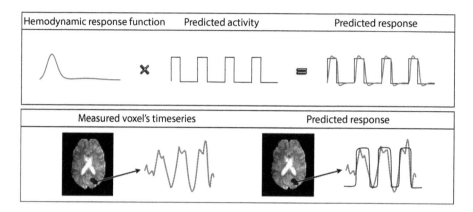

FIGURE 7.5 Two-step fMRI data processing. First step: Convolution of the predicted activity curve of the fMRI experiment with a hemodynamic response function (HRF), producing the so-called *predicted response*. Second step: BOLD signals that match the predicted response (that is the modeled change in BOLD signal) are identified as activation related to stimulus.

adequately. Finally, all volumes are combined to form a 4D volume (otherwise known as the "run"). This 4D volume, or run, can be considered the starting point for analysis. The first part of this analysis is usually referred to as fMRI preprocessing.

7.3.2 Data Preprocessing

Before entering the statistical analysis, it has to be ensured that the data is artifact and noise free. Hence the preprocessing steps involve (1) slice scan timing correction, (2) head motion correction, (3) distortion correction, and (4) spatial and temporal smoothing of the data.

7.3.2.1 Slice-Scan Timing Correction

The MR scanner acquires individual slices within the brain volume at different time intervals; hence the slices represent brain activity at different time-points within a functional volume measurement. This temporal mix up can obviously complicate analysis. For example, if 40 slices are acquired with a volume TR of 3 s, the last slice is measured almost 3 s later than the first slice. Therefore, a timing correction should be applied to align all slices to the same time point reference. In order to accomplish that, time series of individual slices are temporally "shifted" to match the time point reference, assuming the time course of a voxel is smooth when plotted as a dotted line. The optimum degree of temporal shift of the time courses is ensured by resampling the original data accordingly.

7.3.2.2 Head Motion Correction

Head motion correction is the second most common preprocessing step. As in any other MRI acquisition the subject's head should be as stable as possible, but even the most motivated volunteers will slightly move during the scan, affecting the quality of data. Not only the information of a given voxel will be wrongly appointed to another, but moreover the homogeneity of the magnetic field to which the scans where originally optimized will be distorted. A practical cut-off value for rejection of data from further analysis is head movement of more than 5 mm (i.e., 1 or 2 voxels). Any displacement of rigid bodies can be sufficiently described by six parameters as transpired from rotations and translations (displacements) around and along the three orthogonal axes. In that sense, motion correction can be applied by using a reference functional volume as a target volume, to which all other volumes should be realigned. An iterative algorithm is applied and the six parameters adjustment concludes when no further improvement is applicable.

7.3.2.3 Distortion Correction

Distortion correction refers to the correction of field inhomogeneities causing signal dropouts and geometric distortions, especially in regions of the brain with high susceptibility (e.g., temporal lobes, air cavities, etc.). One method used in general to mitigate distortions, is to create a field map of the main field by acquiring two images with differing echo times. Then this map can be retrospectively applied to unwarp EPI distortions using optimized sequence parameters (Weiskopf et al., 2006). Unfortunately, there is no method that could completely remove geometric distortions and signal dropouts, but the field maps solution is very promising and there has been a lot of effort toward this direction (Cusack and Papadakis, 2002; Hutton et al., 2002; Jenkinson, 2003). Field maps only take about a minute to acquire and have the same positioning and image dimensions as the acquired fMRI data. Undistortion can be accomplished with several free tools like FUGUE (FSL) or the Fieldmap Toolbox (SPM).

7.3.2.4 Spatial and Temporal Smoothing

Temporal filtering is the removal of high-frequency signal fluctuations, which are of no interest and are considered noise. Different filters can be used to mitigate this problem. A high-pass filter can be used to remove the lower frequencies, like the reciprocal of twice the TR which is the lowest frequency that can be identified. A low-pass filter can be used to remove higher frequencies, while a band-pass filter can be used to remove all frequencies except the particular range of interest. Temporal smoothing increases the SNR, but may distort temporally relevant parameters of event-related responses. On the other hand, a way to further enhance SNR is spatial smoothing. This is the procedure of averaging the intensities of nearby voxels, thus producing a smooth spatial map of intensity change across the brain or specific region of interest. This averaging is usually done by convolution with a 3D Gaussian kernel, which determines the weights of neighboring voxels by their distance at every spatial point, with the weights decreasing exponentially following a bell-curve distribution (Huettel et al., 2009).

7.3.3 Statistical Analysis

It should be evident by now that even a standard fMRI study gives rise to massive amounts of noisy data with a complicated spatio-temporal correlation structure, and therefore statistical analysis should play a crucial role in understanding the nature of the data and obtaining relevant and robust results (Lindquist, 2008).

> Why?

Because the basic aim of fMRI is to identify the regions of the brain that exhibit different responses in specific stimuli or tasks as compared to control or rest conditions. The presence of physiological noise, distortions and the nature of the signal itself, which is quite weak, makes the identification of the response a challenging task. Statistical analysis can be used as a powerful tool to help differentiate responses from fluctuations, this way protecting from erroneous results and wrong hypotheses. This can be simple correlation analysis or more advanced modeling of the expected hemodynamic response to the stimulation. Several possible statistical corrections can be applied, producing a statistical map that indicates the brain's activation points only in response to the stimulus.

There exist several methods that can be used for the statistical analysis of fMRI data; nevertheless, the most common approach is to consider each voxel separately within the framework of the general linear model (GLM) by fitting the data to a derived model. The GLM is mathematically identical to a multiple regression analysis but stresses its suitability for both multiple qualitative and multiple quantitative variables (Goebel, 2015).

According to this model, the calculated hemodynamic response (HDR) at every time point is equal to the scaled and summed events active at that point. Then, a design matrix is created specifying the active events per time-point. The design matrix and the shape of the HDR are used to generate the predicted voxel response at every time-point using the mathematical procedure of convolution, as this is described in Figure 7.5.

In this basic model, the observed HDR and the predicted HDR are inserted in a scaling procedure using weighting for each event without noise reduction. Then a set of linear equations with more equations than unknowns is generated, producing an exact solution. Overall, the GLM model aims to optimize the scaling weights that minimize the sum of the squares of the error. During the last decade, there has been great progress made in probing brain activity, as well as advances regarding the accuracy and precision of statistical data analysis. A more analytical approach is out of the scope of this book. For more details please refer to the excellent work in the chapters of Rainer Goebel (2015) and Huettel et al. (2009).

7.4 Pre-Surgical Planning with fMRI

Surgical treatment of primary brain tumors is considered successful when the complete removal of the pathology is achieved, without any risk of inducing permanent neurological deficits. In that sense, the applied surgical resection margin should be as accurate as possible without violating functionally eloquent cortical areas. Before the era of fMRI as a non-invasive tool for the visualization of brain function, the functional mapping of brain areas was accomplished by invasive methods such as intraoperative cortical stimulation, implantation of subdural grids, etc. (Sunaert 2006; Vlieger et al., 2004). These techniques are very accurate and effective, but obviously are limited by their difficulty and by the fact that they constitute a surgical procedure with all the limitations of the operating room.

In contrast, fMRI is completely non-invasive, stress-free for the patient and most importantly it can be obtained preoperatively allowing for a well-designed pre-surgical planning. Especially when combined with 3D neuro-navigation tools, it may enable surgeons to visualize the anatomy of a patient's brain during surgery and precisely track the location of their surgical instruments in relation to the specific anatomy, planning the safest route and removal of the lesion (Orringer et al., 2012). Therefore, pre-surgical fMRI offers multiple benefits. First, it gives the opportunity of selection of patients for invasive intraoperative mapping. Second, it gives a non-invasive assessment of neurological deficit that follows a surgical procedure. And third, it may safely guide the surgical procedure itself.

7.5 Resting State fMRI

The human brain cannot completely shut-down, fortunately! A basic level of activity is present even in the absence of any external prompted task or stimulus, and a network of spatially distributed regions that continuously communicate with each other and share information is always active. In fMRI, this activity appears as low-frequency-fluctuations of the BOLD signal, and has been named resting-state fMRI (RS fMRI). Interestingly, as with many scientific findings, the discovery of RS fMRI, was actually made by accident by Biswal et al. (1995), when trying to isolate physiological noise in fMRI data from subjects at rest. They observed that there was a remaining low-frequency signal (<0.1 Hz) after the removal of noise, and in their investigation, they concluded that a high level of correlation existed between the right primary sensorimotor cortex and other motor areas. They thus hypothesized that this was the result of functional connections between these brain regions. More studies followed in order to validate the initial hypothesis and indeed proved that these spontaneous low-frequency fluctuations were blood-oxygenation dependent, like the BOLD signal (Biswal et al., 1997), implying the existence of a network of functional connectivity among different regions of the brain. Moving forward, several other studies provided evidence that RS fMRI has a physiological basis, demonstrating a link between physiological and hemodynamic related BOLD processes (Kenet et al., 2003; Lowe et al., 2000; Mantini et al., 2007).

RS fMRI is undoubtedly a relatively new concept and is in need of appropriate signal collection and analysis methods, with many groups currently working toward this direction with many different approaches, such as seed methods (Fransson, 2005; Song et al., 2008), independent component analysis (Beckmann et al., 2005) and clustering (Thirion et al., 2006; Van den Heuvel et al., 2008; Van den Heuvel and Pol, 2010).

In any case it is evident that RS fMRI can become an invaluable research tool for investigating human brain function (normal and pathological), and that these new examination tools of functional connectivity may further explore diseases, such as schizophrenia, Alzheimer's disease, dementia, and multiple sclerosis, revealing the underlying connectivity and linked pathophysiology.

7.5.1 Resting State fMRI Procedure

The procedure of RS fMRI for the subject is relatively easy, and in any case non-demanding, compared to task fMRI since only remaining calm inside the MRI scanner for about 10 minutes is required, trying not to think anything in particular. Regarding the eyes, whether they should be open or closed, there is a controversy in the literature, with studies showing that when the eyes are open the functional connections between the thalamus and the visual cortex are stronger as opposed to closed (Zou et al., 2009). Nevertheless, it is a matter of planning and an experimental question since there might be a benefit in using the staring at a fixation point technique as well (Tan et al., 2013; Song et al., 2015).

Despite the ease of data collection of RS fMRI, which makes it an attractive and quite popular imaging method, the processing remains challenging. After the collection of RS fMRI data, a pre-processing pipeline is also needed, as in the case of task fMRI, including: spatial and temporal smoothing, motion correction, spatial normalization, etc.

Taking a step further into the understanding of the mechanisms of functional connectivity, current studies are oriented toward the combination of RS fMRI with diffusion tensor imaging (DTI) in order to investigate structural connectivity by evaluating the white matter tract integrity. In other words, research groups are trying to provide insight into how function and structural architecture are related to the human brain. Van den Heuvel and colleagues suggested the existence of structural white matter connections between the functionally linked regions of resting-state networks (van den Heuvel et al., 2009) and other studies have shown a strong correlation between structural and functional connectivity in the brain on a whole-brain scale (Skudlarski et al., 2008) as well as for individual functional networks (Greicius et al., 2009).

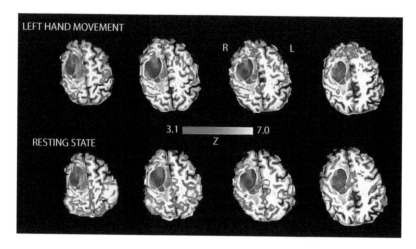

FIGURE 7.6   Motor task fMRI and resting-state fMRI data from a pre-surgical patient with a right frontal lobe glioma. (From Goodyear, B. et al., In In T. D. Papageorgiou et al. [Eds.], *Advanced Brain Neuroimaging Topics in Health and Disease—Methods and Applications,* 2014.)

An example of a typical case of pre-surgical investigation using a combination of fMRI and RS fMRI is shown in Figure 7.6.

A patient with a right-hemisphere frontal lobe glioma in proximity to the motor cortex was investigated with fMRI with the question being whether the patient exhibited a normal pattern of predominantly right-hemisphere motor activity in response to left hand movements and whether this activity was in close proximity or abutting the glioma (Goodyear et al., 2014). Finger tapping at a self-regulated pace was selected as the motor task since this can be performed easily by most patients and is effective in reliably activating sensorimotor regions. Following fMRI, the patient also underwent a 7-minute resting-state scan with eyes open staring at a fixation cross. This clinical fMRI study concluded that the patient exhibited a normal pattern of motor and sensorimotor activity, with an atypical distribution of bilateral premotor activity, possibly the result of functional compensation in response to the impinging glioma. This case demonstrates the aid to pre-surgical planning since it was advised that any resection should attempt to avoid the premotor regions lateral to the glioma.

7.6 Conclusion and the Future of fMRI

Although well established, fMRI is still in its infancy. Nevertheless, there are a growing number of clinical applications in diagnosis or in therapy guidance and development, firmly establishing a growing part of clinical practice. Table 7.2 illustrates the established clinical applications of fMRI, accompanied by those that are expected to be established soon, and the ones that are expected to emerge in the near future.

It is true that the field of fMRI, including paradigm selection, data acquisition and analysis, is indeed quite complex, and in that sense, the prediction of its future applications is extremely difficult. It is also true that a rather big part of fMRI research belongs to the domain called "cognitive neuroscience," typically meaning the exploration of the way of thinking and of behavioral aspects, also expanding into other aspects of experimental social sciences such as social neurosciences and neuroeconomics. Nevertheless, it is certain that regarding the evaluation

TABLE 7.2   Established Clinical Applications of fMRI, Accompanied by Those Which Are Expected to Be Established Soon, and the Ones That Are Expected to Emerge in the Near Future

Established clinical applications of fMRI	Pre-surgical mapping for neurosurgical approaches
	Lateralization for temporal lobe epilepsy (TLE) surgery
	Mapping of ictal foci in patients with focal epilepsy
	Resting state fMRI for cognitive impairment
	Post-surgical brain evaluation
	Study of neurologic disorders
Near future clinical applications of fMRI	Evaluation of brain's plasticity in stroke
	Pharmacological fMRI
	Prediction of patient benefits
	Chronic pain evaluation and management
Future clinical applications of fMRI	Real-time fMRI
	Development and evaluation of new treatment strategies

and understanding of brain functionality and neural mechanisms, fMRI will continue to be the leading neuroscience technique.

References

Bandettini, P. A., Jesmanowicz, A., Wong, E. C., and Hyde, J. S. (1993). Processing strategies for time-course data sets in functional mri of the human brain. *Magnetic Resonance in Medicine*, 30(2), 161–173. doi:10.1002/mrm.1910300204

Beckmann, C. F., DeLuca, M., Devlin, J. T., and Smith, S. M. (2005). Investigations into resting-state connectivity using independent component analysis. Philosophical transactions of the Royal Society of London. *Series B, Biological Sciences*, 360(1457), 1001–1013.

Biswal, B., Haughton, V. M., Hyde, J. S., and Yetkin, F. Z. (1995). Functional connectivity in the motor cortex of resting human brain using echo-planar MRI. *Magnetic Resonance in Medicine*, 34(4), 537–541.

Biswal, B. B., Kylen, J. V., and Hyde, J. S. (1997). Simultaneous assessment of flow and BOLD signals in resting-state functional connectivity maps. *NMR in Biomedicine*, 10(4–5), 165–170. doi:10.1002/(sici)1099-1492(199706/08)10:4/5<165::aid-nbm454>3.0.co;2-7

Boynton, G. M., Engel, S. A., Glover, G. H., and Heeger, D. J. (1996). Linear systems analysis of functional magnetic resonance imaging in human V1. *The Journal of Neuroscience: The Official Journal of the Society for Neuroscience*, 16(13), 4207–4221.

Chen, N. and Wyrwicz, A. M. (2001). Optimized distortion correction technique for echo planarimaging. *Magnetic Resonance in Medicine*, 45(3), 525–528. doi:10.1002/1522-2594(200103)45:3<525::aid-mrm1070>3.0.co;2-s

Cusack, R. and Papadakis, N. (2002). New robust 3-D phase unwrapping algorithms: Application to magnetic field mapping and undistorting echoplanar images. *NeuroImage*, 16(3), 754–764. doi:10.1006/nimg.2002.1092

Dennis, N. A., Daselaar, S., and Cabeza, R. (2007). Effects of aging on transient and sustained successful memory encoding activity. *Neurobiology of Aging*, 28(11), 1749–1758. doi:10.1016/j.neurobiolaging.2006.07.006

Dosenbach, N. U., Fair, D. A., Cohen, A. L., Schlaggar, B. L., and Petersen, S. E. (2008). A dual-networks architecture of top-down control. *Trends in Cognitive Sciences*, 12(3), 99–105. doi:10.1016/j.tics.2008.01.001

Fox, M. D., Snyder, A. Z., Barch, D. M., Gusnard, D. A., and Raichle, M. E. (2005). Transient BOLD responses at block transitions. *NeuroImage*, 28(4), 956–966. doi:10.1016/j.neuroimage.2005.06.025

Fransson, P. (2005). Spontaneous low-frequency BOLD signal fluctuations: An fMRI investigation of the resting-state default mode of brain function hypothesis. *Human Brain Mapping, 26*(1), 15–29. doi:10.1002/hbm.20113

Goebel, R. (2015). Analysis of Functional MRI Data. In K. Uludağ et al. (Eds.), *fMRI: From Nuclear Spins to Brain Functions*, New York, NY: Springer.

Goodyear, B., Liebenthal, E., and Mosher, V. (2014). Active and passive fMRI for presurgical mapping of motor and language cortex. In T. D. Papageorgiou, G, I. Christopoulos and S. M. Smirnakis (Eds.), *Advanced Brain Neuroimaging Topics in Health and Disease—Methods and Applications*, Rijeka, Croatia: INTECH. doi:10.5772/58269

Gore, J. C. (2003). Principles and practice of functional MRI of the human brain. *Journal of Clinical Investigation, 112*(1), 4–9. doi:10.1172/jci19010

Greicius, M. D., Supekar, K., Menon, V., and Dougherty, R. F. (2009). Resting-state functional connectivity reflects structural connectivity in the default mode network. *Cerebral Cortex, 19*(1), 72–78. doi:10.1093/cercor/bhn059

Griswold, M. A., Heidemann, R. M., Haase, A., Jakob, P.M., Jellus, V., Kiefer, B. et al. (2002). Generalized autocalibrating partially parallel acquisitions (GRAPPA). *Magnetic Resonance in Medicine, 47*(6), 1202–1210.

Huettel, S. A., Song, A. W., and McCarthy, G. (2009). *Functional Magnetic Resonance Imaging*. Sunderland, Massachusetts: Sinauer Associates, Inc.

Hutton, C., Bork, A., Josephs, O., Deichmann, R., Ashburner, J., and Turner, R. (2002). Image distortion correction in fMRI: A quantitative evaluation. *NeuroImage, 16*(1), 217–240. doi:10.1006/nimg.2001.1054

Jenkinson, M. (2003). Fast, automated, N-dimensional phase-unwrapping algorithm. *Magnetic Resonance in Medicine, 49*(1), 193–197. doi:10.1002/mrm.10354

Jezzard, P. and Balaban, R. S. (1995). Correction for geometric distortion in echo planar images from B0 field variations. *Magnetic Resonance in Medicine, 34*(1), 65–73. doi:10.1002/mrm.1910340111

Jezzard, P. and Ramsay, N. F. (2004). Functional MRI. In P. Tofts (Ed.), *Quantitative MRI of the Brain: Measuring Changes Caused by Disease*. Chichester, UK: John Wiley & Sons.

Kenet, T., Arieli, A., Bibitchkov, D., Grinvald, A., and Tsodyks, M. (2003). Spontaneously emerging cortical representations of visual attributes. *Nature, 425*(6961), 954–956.

Konishi, S., Yoneyama, R., Itagaki, H., Uchida, I., Nakajima, K., Kato, H. et al. (1996). Transient brain activity used in magnetic resonance imaging to detect functional areas. *NeuroReport, 8*(1), 19–23. doi:10.1097/00001756-199612200-00005

Lindquist, M. A. (2008). The statistical analysis of fMRI data. *Statistical Science, 23*(4), 439–464. doi:10.1214/09-sts282

Liu, G., Sobering, G., Duyn, J., and Moonen, C. T. (1993). A. functional MRI technique combining principles of echo-shifting with a train of observations (PRESTO). *Magnetic Resonance in Medicine, 30*(6), 764-768. doi:10.1002/mrm.1910300617

Liu, G., Sobering, G., Olson, A. W., Gelderen, P. V., and Moonen, C. T. (1993). Fast echo-shifted gradient-recalled MRI: Combining a short repetition time with variable T2* weighting. *Magnetic Resonance in Medicine, 30*(1), 68-75. doi:10.1002/mrm.1910300111

Lowe, M. J., Dzemidzic, M., Lurito, J. T., Mathews, V. P., and Phillips, M. D. (2000). Correlations in low-frequency BOLD fluctuations reflect cortico-cortical connections. *NeuroImage, 12*(5), 582–587. doi:10.1006/nimg.2000.0654

Mantini, D., Corbetta, M., Gratta, C. D., Perrucci, M. G., and Romani, G. L. (2007). Electrophy-siological signatures of resting state networks in the human brain. Proceedings of the National Academy of Sciences of the United States of America, *104*(32), 13170–13175.

Maus, B., Breukelen, G. V., Goebel, R., and Berger, M. (2010). Optimization of blocked designs in fMRI studies. *NeuroImage, 47*, 373–390. doi:10.1016/s1053-8119(09)71198-x

Meltzer, J. A., Negishi, M., and Constable, R. T. (2008). Biphasic hemodynamic responses influence deactivation and may mask activation in block-design fMRI paradigms. *Human Brain Mapping, 29*(4), 385–399. doi:10.1002/hbm.20391

Menon, R. S., Ogawa, S., Tank, D. W., and Uğurbil, K. (1993). Tesla gradient recalled echo characteristics of photic stimulation-induced signal changes in the human primary visual cortex. *Magnetic Resonance in Medicine, 30*(3), 380–386. doi:10.1002/mrm.1910300317

Mosso, A. (1881). *Ueber den Kreislauf des Blutes im Menschlichen Gehirn.* Leipzig, Germany: Verlag von Veit & Company.

Norris, D. G. (2015). *Pulse sequences for fMRI.* In K. Uludağ, K. Uğurbil, and L. Berliner (Eds.), *fMRI: From Nuclear Spins to Brain Functions,* New York, NY: Springer.

Ogawa, S. and Lee, T. (1990). Magnetic resonance imaging of blood vessels at high fields: In vivo and in vitro measurements and image simulation. *Magnetic Resonance in Medicine, 16*(1), 9–18. doi:10.1002/mrm.1910160103

Ogawa, S., Glynn, P., Lee, T. M., and Nayak, A. S. (1990). Oxygenation-sensitive contrast in magnetic resonance image of rodent brain at high magnetic fields. *Magnetic Resonance in Medicine, 14*(1), 68–78.

Orringer, D. A., Golby, A., and Jolesz, F. (2012). Neuronavigation in the surgical management of brain tumors: Current and future trends. *Expert Review of Medical Devices, 9*(5), 491–500. doi:10.1586/erd.12.42

Petersen, S. E. and Dubis, J. W. (2012). The mixed block/event-related design. *NeuroImage, 62*(2), 1177–1184. doi:10.1016/j.neuroimage.2011.09.084

Poser, B. A. and Norris, D. G. (2007). Fast spin echo sequences for BOLD functional MRI. *Magnetic Resonance Materials in Physics, Biology and Medicine, 20*(1), 11–17. doi:10.1007/s10334-006-0063-x

Pruessmann, K. P., Weiger, M., Scheidegger, M. B., and Boesiger, P. (1999). SENSE: Sensitivity encoding for fast MRI. *Magnetic Resonance in Medicine, 42*(5), 952–962. doi:10.1002/(sici)1522-2594(199911)42:5<952::aid-mrm16>3.3.co;2-j

Skudlarski, P., Jagannathan, K., Calhoun, V. D., Hampson, M., Skudlarska, B. A., and Pearlson, G. (2008). Measuring brain connectivity: Diffusion tensor imaging validates resting state temporal correlations. *NeuroImage, 43*(3), 554–561. doi:10.1016/j.neuroimage.2008.07.063

Song, M., Jiang, T., Li, J., Liu, Y., Tian, L., Yu, C., and Zhou, Y. (2008). Brain spontaneous functional connectivity and intelligence. *NeuroImage, 41*(3), 1168–1176.

Song, X., Zhou, S., Zhang, Y., Liu, Y., Zhu, H., and Gao, J. H. (2015). Frequency-dependent modulation of regional synchrony in the human brain by eyes open and eyes closed resting-states. *PLOS One, 10*(11), e0141507. doi:10.1371/journal.pone.0141507

Sunaert, S. (2006). Presurgical planning for tumor resectioning. *Journal of Magnetic Resonance Imaging, 23*(6), 887–905.

Tan, B., Kong, X., Yang, P., Jin, Z., and Li, L. (2013). The difference of brain functional connectivity between eyes-closed and eyes-open using graph theoretical analysis. *Computational and Mathematical Methods in Medicine, 2013,* 1–15. doi:10.1155/2013/976365

Taylor, S. F., Stern, E. R., and Gehring, W. J. (2007). Neural systems for error monitoring. *The Neuroscientist, 13*(2), 160–172. doi:10.1177/1073858406298184

Thirion, B., Dodel, S., and Poline, J. (2006). Detection of signal synchronizations in resting-state fMRI datasets. *NeuroImage, 29*(1), 321–327. doi:10.1016/j.neuroimage.2005.06.054

Uğurbil, K., Ellermann, J., Garwood, M., Hendrich, K., Hinke, R., Hu, X. et al. (1993). *Imaging at high magnetic fields: Initial experiences at 4 T. Magnetic Resonance Quarterly, 9*(4), 259–277.

Van den Heuvel, M. P., and Pol, H. E. (2010). Exploring the brain network: A review on resting-state fMRI functional connectivity. *European Neuropsychopharmacology, 20*(8), 519–534. doi:10.1016/j.euroneuro.2010.03.008

Van den Heuvel, M. P., Mandl, R. C., Kahn, R. S., and Pol, H. E. (2009). Functionally linked resting-state networks reflect the underlying structural connectivity architecture of the human brain. *Human Brain Mapping, 30*(10), 3127–3141. doi:10.1002/hbm.20737

Van den Heuvel, M. V., Mandl, R., and Pol, H. H. (2008). Normalized cut group clustering of resting-state fMRI data. *PLOS One, 3*(4), e2001. doi:10.1371/journal.pone.0002001

Velanova, K., Buckner, R. L., Jacoby, L. L., McAvoy, M. P., Petersen, S. E., and Wheeler, M. E. (2003). Functional-anatomic correlates of sustained and transient processing components engaged during controlled retrieval. *The Journal of Neuroscience: The Official Journal of the Society for Neuroscience, 23*(24), 8460–8470.

Visscher, K. M., Miezin, F. M., Kelly, J. E., Buckner, R. L., Donaldson, D. I., Mcavoy, M. P., and Petersen, S. E. (2003). Mixed blocked/event-related designs separate transient and sustained activity in fMRI. *NeuroImage, 19*(4), 1694–1708. doi:10.1016/s1053-8119(03)00178-2

Vlieger, E., Majoie, C. B., Leenstra, S., and Heeten, G. J. (2004). Functional magnetic resonance imaging for neurosurgical planning in neurooncology. *European Radiology, 14*(7), 1143–1153. doi:10.1007/s00330-004-2328-y

Weiskopf, N., Hutton, C., Josephs, O., and Deichmann, R. (2006). Optimal EPI parameters for reduction of susceptibility-induced BOLD sensitivity losses: A whole-brain analysis at 3 T and 1.5 T. *NeuroImage, 33*(2), 493–504. doi:10.1016/j.neuroimage.2006.07.029

Weiskopf, N., Klose, U., Birbaumer, N., and Mathiak, K. (2005). Single-shot compensation of image distortions and BOLD contrast optimization using multi-echo EPI for real-time fMRI. *NeuroImage, 24*(4), 1068–1079. doi:10.1016/j.neuroimage.2004.10.012

Yacoub, E., Moortele, P. V., Shmuel, A., and Uğurbil, K. (2005). Signal and noise characteristics of Hahn SE and GE BOLD fMRI at 7 T in humans. *NeuroImage, 24*(3), 738–750. doi:10.1016/j.neuroimage.2004.09.002

Ye, Y., Zhuo, Y., Xue, R., and Zhou, X. J. (2010). BOLD fMRI using a modified HASTE sequence. *NeuroImage, 49*(1), 457–466. doi:10.1016/j.neuroimage.2009.07.044

Zaitsev, M., Hennig, J., and Speck, O. (2004). Point spread function mapping with parallel imaging techniques and high acceleration factors: Fast, robust, and flexible method for echo-planar imaging distortion correction. *Magnetic Resonance in Medicine, 52*(5), 1156–1166. doi:10.1002/mrm.20261

Zou, Q., Long, X., Zuo, X., Yan, C., Zhu, C., Yang, Y. et al. (2009). Functional connectivity between the thalamus and visual cortex under eyes closed and eyes open conditions: A resting-state fMRI study. *Human Brain Mapping, 30*(9), 3066–3078. doi:10.1002/hbm.2072

Artifacts and Pitfalls of fMRI

8.1 Introduction to Quantitative fMRI Limitations

Focus Point

- fMRI is an indirect measurement of neural activity, based on hemodynamic changes.
- Causation of brain stimulus remains debatable.
- fMRI signal is weak and embedded in a noise contaminated environment.
- fMRI is affected by a series of technical issues limiting its widespread clinical use.
- fMRI is especially sensitive to fast imaging artifacts.
- The major concern in BOLD fMRI is the localization uncertainty of neural activity.
- Physiological noise is an issue.

Functional magnetic resonance imaging (fMRI) is, in principle, a technique for the spatial localization of changes in image intensity of the brain, following the hemodynamic response of a certain experimental task or stimulus. In fact, these stimulus-related intensity changes may as well be only a small percentage of the overall collected signal, which is by default embedded in a noise contaminated environment by several other parameters that may affect image acquisition. In that sense, it is evident that there are several limitations associated with the data collection as well as the interpretation of the acquired signals that have to be taken into account. Moreover, despite being directly correlated with neuron's electrical activity, as explained in the previous chapter, fMRI is in fact an indirect evaluation method. Therefore, although studies have been able to demonstrate brain to stimulus correlation, causation is still a major issue. In conclusion, it seems that fMRI is a very powerful and continuously evolving neuroimaging technique, but adding to other limitations it can be sometimes misused and over-interpreted.

Overall, the limitations could be summed in three distinct categories:

- Image acquisition limitations
- Physiological noise and motion limitations
- Interpretation limitations

8.2 Image Acquisition Limitations

8.2.1 Spatial and Temporal Resolution

One of the pre-requirements of the fMRI technique is the capability of performing high-speed MRI, so that the acquired image is effectively "frozen" during the acquisitions, contributing to the elimination of artifacts from physiological processes. Most clinical and research fMRI studies use very fast, sensitive to the BOLD phenomenon, pulse sequences, in order to have adequate spatial (about 2–3 mm), and temporal (about 2–3 sec) resolution. Thus, a two-dimensional $T2^*$-weighted gradient echo (GRE) sequence with the well-known echo planar imaging (EPI) readout is usually used (Norris, 2006). Unfortunately, as analytically discussed in Chapter 2, EPI is generally vulnerable to susceptibility effects.

Why?

Because, the magnetic susceptibility difference of multiple tissues that may be contained in a voxel under investigation can produce macroscopic magnetic field gradients causing intravoxel dephasing. Hence, intravoxel spins may experience different magnetic fields, thus precessing at different frequencies resulting in a signal loss due to dephasing.

It follows that this phenomenon will be most prominent in regions of high susceptibility differences, like bone and air boundaries, which unfortunately include areas with great clinical interest in fMRI, such as the orbitofrontal cortex or the medial temporal and the inferior temporal lobes. These areas are important in visual and cognitive processing, including language and memory functions. A signal loss in these regions can reduce the fMRI activation sensitivity, and must be taken into account since they may be neglected when the statistical maps are overlaid on high resolution T1-weighted anatomic images without susceptibility artifacts.

Signal loss can be studied by simulating the effect of field gradients on the EPI signal, and it has been shown (using field maps), that it depends on the image orientation, echo time (TE), and spatial resolution (Ojemann et al., 1997). These acquired field maps can also be used to correct such distortions (Jezzard and Balaban, 1995). Figure 8.1 illustrates an acquired field map along with the reference anatomical image and shows how the choice of phase encoding direction can affect localization accuracy when fieldmap-based distortion compensation techniques are used during data analysis (Olman et al., 2009).

In clinical fMRI, susceptibility artifacts are not limited to anatomical inhomogeneities but can be induced by a number of sources usually related to previous surgery, including clips, stent grafts, etc. Surgery-related materials can induce strong macroscopic field gradients generating appreciable signal losses. Previous surgery can also induce hemorrhages, which can in turn cause susceptibility artifacts, which may degrade pre-surgical fMRI mapping, although in most cases it may still be possible (Peck et al., 2009), keeping in mind that findings should be interpreted with caution in the presence of post-surgical artifacts.

As mentioned earlier, the larger the voxel size, the faster the signal dispersion or signal loss. Hence, fMRI's spatial resolution is of paramount importance because it is the reduction of voxel size that can reduce the susceptibility effects. Of course, the reduction of voxel size comes at the cost of reduced brain coverage and the question regarding temporal resolution since the dephasing increases with time. In clinical routine, there is a constant tradeoff between spatial and temporal resolution with optimum BOLD phenomenon sensitivity that needs to be accomplished. Echo planar images have an inferior spatial resolution and overall image quality

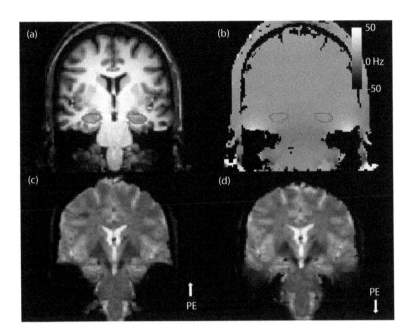

FIGURE 8.1 (a) Reference ANATOMICAL image for an oblique coronal slice (perpendicular to the hippocampal axis) through the anterior hippocampus (line boundary). (b) Field map (bar indicating frequency offsets in Hz) for the same slice; the ROI boundary is the same as in (a). (c) High resolution, full field of view EPI image with phase-encode (PE) in the foot–head direction. Susceptibility artifacts decrease the static magnetic field in the medial temporal lobe and shift the signal from the PHG toward the feet (notably, the hippocampus is not affected). Increased B0 (static magnetic field strength) in the lateral temporal lobe shifts the signal toward the top of the head. (d) Reversing the phase-encode direction reverses the direction of distortion. The choice of phase-encode direction can affect localization accuracy when field-map-based distortion compensation techniques are used during data analysis. (From Olman, C. A. et al., *PLOS One*, 4, e8160, 2009.)

than T1 or T2 anatomical scans. The standard maximum resolution of single-shot EPI is about 2 mm². Nevertheless, one of the most promising developments in fMRI scanner technology has been the use of multiple parallel RF receive coils to help spatially encode the data, thus allowing for much higher resolution with a single excitation pulse, which can allow functional image resolutions of about 1 mm³ (Bandettini, 2009).

Interestingly, it is not the technical limits of the acquisition methods that determine the maximum spatial resolution of the fMRI procedures, but rather the relatively wide spatial spread of the oxygenation and perfusion changes following brain activation, or the so-called "*hemodynamic point spread function.*" It has been shown that the *hemodynamic point spread function* of the $T2^*$-weighted gradient echo BOLD response can be of the order of ~3.5 mm at 1.5 T (Engel, 1997), and less than 2 mm at 7 T (Shmuel et al., 2007). Shmuel's group has also suggested that the point-spread function of T2 and $T2^*$ BOLD responses at 7 T relative to metabolic activity have been estimated as ~0.8 mm and ~1.0 mm, respectively (Shmuel and Maier, 2015). Obviously, for a better spatial resolution, the higher the magnetic field the greater the image signal to noise, and the larger the functional contrast, allowing for higher signal per voxel volume.

On the other hand, it has to be mentioned at this point that most fMRI studies involve several techniques and analysis procedures in their data analysis pipeline, that effectively reduce the spatial resolution to about 10 mm³, probably making all the aforementioned efforts for a spatial resolution of less than 2 mm³ potentially redundant. The aforementioned analysis of fMRI data involves multiple stages of data pre-processing before the activation can be statistically detected, such as spatial smoothing, spatial normalization, and multi-subject averaging (Mikl et al., 2008). In other words, high spatial resolution is only absolutely necessary and beneficial when single-subject assessment is involved, where the advantages of collecting high resolution data would yield more subtle and patient specific information.

The hemodynamic response also poses limitations in temporal resolution of the fMRI signals acquired. EPI images have an acquisition window of about 20–30 ms, which relative to the inertia and variability of the hemodynamic response is quite fast and adequate. Nevertheless, the problem arises when the signal is influenced by the underlying vasculature a voxel covers. That is, if a voxel happens to cover large vessel effects, the magnitude of the signal can be up to an order of magnitude larger than the capillary effects and the timing somewhat delayed for up to 4 sec. It follows that signal temporal dynamics varies, and it can generally be described by an increase of the fMRI signal, approximately 2 sec following neuronal activity as well as a plateau in the so-called "on" state for about 7–10 sec (Buxton et al., 2004).

This means that the determination of the precise timing of signal activation between adjacent regions of the brain can be very difficult, and the required temporal resolution to achieve this goal would be of the order of tens of milliseconds.

Several hypotheses have been made regarding the hemodynamic responses of vessels and capillaries, nevertheless the precise mechanisms of these temporal variations are yet to be completely determined. Most hemodynamic response modeling research to date has focused on the magnitude of evoked activation, although timing and shape information should be established first. However, there is a growing interest in measuring onset, peak latency and duration of evoked fMRI responses. A number of fitting procedures exist that potentially allow the characterization of the latency and duration of fMRI responses. Ideally only one model that extracts the shape of the hemodynamic response function (HRF) to different types of cognitive events would be required (Bellgowan et al., 2003; Thompson, Engel, and Olman, 2014). Nevertheless, most groups that analyze functional neuroimaging data assume a fixed shape for the measured response, by regressing data onto a canonical HRF (Smith et al., 2011; Thompson et al., 2014). This approach is widely used, but obviously has limitations because it is unknown whether the timing of fMRI responses changes as neural activity increases (Li et al., 2008; Thompson et al., 2014).

8.2.2 Spatial and Temporal fMRI Resolution—Mitigating Strategies

There are several approaches to increasing fMRI spatial resolution. The most obvious one, as mentioned earlier, is to increase the magnetic field strength, thus increasing the functional signal to noise producing useful data in clinically acceptable times (Murphy et al., 2007). Hence, it is generally recommended and preferable to perform fMRI at 3 T or higher. Especially regarding individual subject assessment, which is by default not spatially smoothed, normalized or averaged, fMRI should only be performed at high magnetic field strengths (≥3 T).

Regarding the mitigation of the hemodynamic point spread function, besides the field strength increase, the primary effort has been to reduce the effects of large vessels, mainly by using spin-echo imaging. Previous evidence showed that, due to refocusing of static dephasing effects around large vessels, spin-echo (SE) BOLD signals offer an increased linearity and promptness with respect to gradient-echo (GE) acquisition, and have been proposed as a potential alternative to obtain increased functional localization to the capillary bed (Chiacchiaretta and Ferretti, 2015; Jochimsen et al., 2004; Norris, 2012; Parkes et al., 2005). In fact, the 180° refocusing pulse that rephases the contribution around larger vessels is also effective at recovering susceptibility effect's signal loss, making SE sequences superior regarding spatial specificity of the microvasculature but only at high field strengths ($\geq$7 T) (Halai et al., 2014; Schwarzbauer et al., 2010).

Figure 8.2 shows GE and SE seed-based functional connectivity maps for GE and SE sequences obtained from the random effects group analysis showing resting state networks. It is evident that the spatial patterns were largely overlapping for the two sequences (Figure 8.2b), except for the ventromedial prefrontal cortex of the DMN, which did not show a significant connectivity with the seed in the GE acquisition (Chiacchiaretta and Ferretti, 2015).

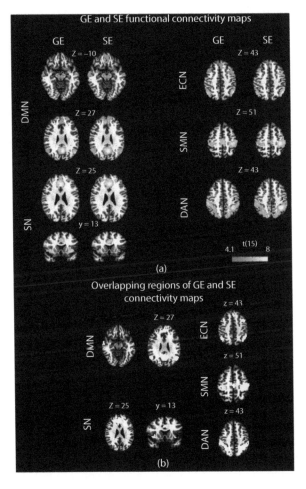

FIGURE 8.2   Seed-based connectivity maps for gradient-echo (GE) and spin-echo (SE) obtained from the random effects group analysis showing the following resting state networks: default mode network (DMN), executive control network (ECN), salience network (SN), dorsal attention network (DAN), and sensorimotor network (SMN) (a). The group statistical maps were thresholded at $p < 0.05$, corrected for multiple comparisons using a cluster-size algorithm, and superimposed on the Talairach template; (b) overlapping regions of GE and SE connectivity maps. (From Chiacchiaretta, P. and Ferretti, A., *PLOS One*, 10, e0120398, 2015.)

Another method to increase spatial resolution is an attempt to calibrate the hemodynamic factors that influence the BOLD signal change. For example, differences in the baseline cerebral blood flow and volume, or the rate of breathing and the heart rate might produce variations in the BOLD response (Birn et al., 2006; Thomason and Glover, 2008). Thus, the goal would be to better discriminate fMRI signal components that are related to neural activity from those that result from intrinsic properties of the local vasculature (Thomason and Glover, 2008).

The proposed spatial calibration methods can be described under the general idea of creating a map of potential BOLD signal magnitudes by providing an evenly distributed hemodynamic brain stress such as hypercapnia (Bandettini and Wong, 1997; Thomason et al., 2007) with CO_2 inhalation stress (Chiarelli et al., 2007) or Breath Holding (BH) techniques (Handwerker et al., 2007; Thomason et al., 2005).

The map is then used for the correction of vascular reactivity-induced effects in the BOLD signal by dividing activation-induced signal changes, by this map of signal change to the global stress. Measurements taken during stress are used to identify individual- and region-specific differences in hemodynamic responsivity, and applying correction for these differences to cognitive paradigms. Very recently CO_2 stress and BH calibration techniques have been used to quantify alterations in regional cerebrovascular responsiveness, by applying a global physiological stimulus as a "task" in an effort to assess the cerebrovascular integrity of the entire brain (Gonzales et al., 2014; Mutch et al., 2014).

Regarding temporal resolution the most direct mitigating strategy would be to first positively identify and then remove large vessel effects by using calibration methods, thus reducing physiologic fluctuations and aiming to temporal signal to noise. The major obstacle in comparing the hemodynamic response across systems, and even within systems, is that different brain regions exhibit biologically determined differences (spatial bias) in hemodynamic response properties that are unrelated to the underlying neural function (Bellgowan et al., 2003).

8.2.3 EPI-Related Image Distortions

The main advantage of EPI sequences is that an entire image can be acquired in a fraction of a second practically "freezing" motion during acquisition. Nevertheless, the main drawback of EPI is its high sensitivity to geometric distortions, especially at the boundaries of heterogeneous areas of the brain. This is because there is no such thing as a purely homogeneous magnetic field because even in a perfectly shimmed magnetic field the human head will magnetize unevenly due to air cavities or bone-tissue boundaries, so that the MR frequency may slightly differ from one point to another. These rather small frequency differences (about 1 ppm) unfortunately result in mislocalization of the signal in the phase encoding direction and signal intensity loss in the resulting images.

This signal mislocalization may be a frequent cause of concern in fMRI as the brain activation map is superimposed or fused onto the higher resolution structural images, resulting in improper registration and information errors.

Generally, however, these artifacts may be considerably reduced by distortion correction methods based on the local point spread function (In and Speck, 2012), application of parallel imaging with faster EPI readout using high-performance gradient systems and by image post-processing. One example of the application of the field map technique, which is one of the most used distortion correction methods, is depicted in Figure 8.3. For an analytical description of

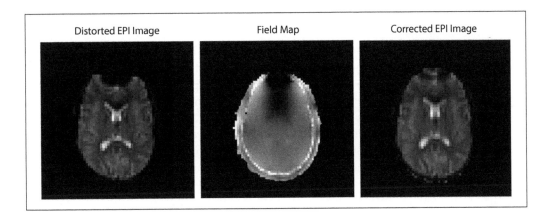

FIGURE 8.3   On the left, the distorted (by field inhomogeneity) EPI image is depicted. From the field map image (middle) we get the magnitude of spatial distortions in the phase encoding direction. Then, the field map image is used to calculate distortion and "unwarp" the EPI image, as shown on the right. (Adapted from FSL. Available at https://fsl.fmrib.ox.ac.uk/fsl/fslwiki/ Copyright © 2000-2017, FMRIB Analysis Group and MGH, Boston.)

distortion correction algorithms, (see Hutton et al., 2002) and the more recent review paper by (Jezzard, 2012).

8.3 Physiological Noise and Motion Limitations

Physiological noise and head motion are two patient-related critically important confounds that usually affect the BOLD signal, degrading sensitivity and specificity of the fMRI outcome. When fMRI is performed in the midbrain, motion is a secondary problem compared to the aforementioned susceptibility effects. But this is true only for pulsatile cranio-caudal motion and not for rigid body motion.

Jezzard (1999), defined physiological noise as signal changes that are caused by the subject's physiology, for example, metabolic and other fluctuations associated with cardiac and respiratory processes, excluding paradigm-related brain activation. The sources of physiological noise in brain fMRI can therefore be identified as both induced by the cardiac and the respiratory cycle. Mechanisms that contribute to noise due to the cardiac cycle are considered to be the changes in cerebral blood flow (CBF) and volume (CBV), arterial pulsatility, and CSF flow (Krüger and Glover, 2001), while mechanisms that contribute to noise due to the respiratory cycle include induced changes in arterial CO_2 partial pressure (Wise et al., 2004), and possible changes in the main magnetic field (B0) (Raj et al., 2001).

These mechanisms affect the fMRI signal in different ways, but generally, bulk motion will lead to similar artifacts as rigid body head motion. For instance, the magnetic field can be altered either by the movement of the subject's chest (Brosch et al., 2002) or by susceptibility induced changes in the lungs (Raj et al., 2001), leading to apparent movement due to the geometric distortion associated with the varying magnetic field. These magnetic field changes usually result in a shift of the MR image in the phase-encoding direction and the displacements of voxels in the image cause artifacts that are particularly harmful near areas with high susceptibility differences, such as at the edge of the brainstem (Brooks et al., 2013). Other mechanisms such as oxygenation or blood volume changes create alterations in local blood susceptibility, and these changes will induce BOLD-related signal variations in the same pattern as the expected brain activations induced by paradigms. Breathing can also affect the head motion, which may spatially alter spin history (Friston et al., 1996) and, along with cardiac pulsation, can cause the brain stem to push up into the surrounding brain tissue. In either case, the main magnetic field is altered by the deformation and cerebrospinal fluid movement (Dagli et al., 1999). There are also ways that the signal can be affected via changes in tissue composition or via inflow effects (Brooks et al., 2013). That is, vessel's pulsations that are generated by cardiac-induced pressure changes may produce small movements in and around large blood vessels (Dagli et al., 1999). In any case, Soellinger et al. (2007) reported a heartbeat-correlated pulsatile cranio-caudal motion of no more than 0.2-mm displacements with peak velocities ≤2 mm/s.

It is very interesting to mention here that all the aforementioned mechanisms of physiological noise are field strength dependent. Moreover, while the signal-to-noise ratio (SNR) increases linearly with field strength, the relative contribution of physiological noise increases with the square of the field (Triantafyllou et al., 2005). Consequently, strategies for reducing the effects of physiological fluctuations are particularly important in fMRI studies at higher field strengths such as 7 T because physiological noise can become the dominant source of noise, giving rise to false positive activations. Especially for subtle BOLD responses, it is of paramount importance to increase the temporal SNR (tSNR) by the use of optimal physiological noise correction methods (Hutton et al., 2011).

Indicatively, an overview of the SNR values that were reported for real fMRI data from a recent study are presented in Figure 8.4. Many authors explicitly reported tSNR values ranging from 4.42 to 280, while in a few other cases, the CNR values that were reported varied from 0.5 to 1.8. Both in the experimental and simulation studies in this literature search, the reported values demonstrated a range that was much wider than can be explained by natural variation only (Welvaert and Rosseel, 2013).

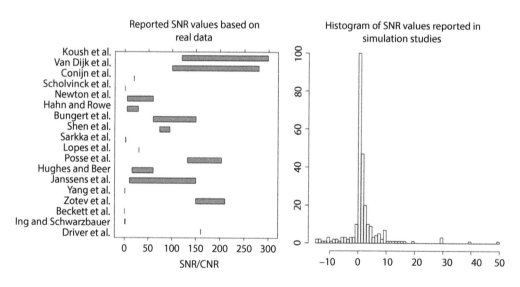

FIGURE 8.4 Overview of reported SNR values in real data (left panel) and simulated data (right panel). (From Welvaert, M. and Rosseel, Y., *PLOS One*, 8, e77089, 2013.)

8.3.1 Physiological Noise—Mitigating Strategies

Several different mitigating strategies have been proposed to minimize the influence of physiological noise in fMRI experiments, broadly falling into the following categories:

1. Gating (cardiac or respiratory), referring to techniques that attempt to effectively "freeze" physiological motion (prospective techniques)
2. Acquisition-based image corrections, referring to techniques aiming to correct for signal intensity variations
3. Calibration, referring to techniques that include additional scans to identify noise sources and retrospectively eliminate them (retrospective techniques)

8.3.1.1 Cardiac Gating

Cardiac pulsations can induce temporal signal fluctuations, and it has been shown that within a cardiac cycle, brain tissue (particularly at the base of the brain) can move up to a range of several millimeters. Cardiac gating ensures that an image is collected at the same phase of each cardiac cycle, typically by phase locking the sampling to a fixed time point during each cardiac cycle ensuring that the images are repeatedly registered, as this is illustrated in Figure 8.5.

The only drawback is that the heart rate can be variable, hence the TR will no longer be "fixed" but it will vary accordingly. A cardiac cycle dependent TR will lead to large signal variations due to varying longitudinal (T1) saturation, meaning that that there will be different amounts of T1 relaxation between samples. This problem has to be accounted for and corrected.

There are two approaches regarding this correction. The first approach is to use an *effective TR* so that the extra amount of partial saturation of the MR signal is corrected (increased or decreased appropriately) by an adjustment of the intermediate time between samples (Malinen et al., 2006). Nevertheless, additional processing is required to minimize induced effects of a variable TR, such as a lack of information on the relative magnitude of cardiac noise, or a difficulty to accurately correct for the signal variations due to the alteration of repetition time.

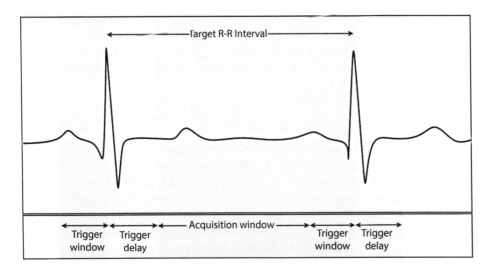

FIGURE 8.5 Cardiac gating ensures that an image is collected at the same phase of the cardiac cycle, typically by phase locking the sampling to a particular time point during each cardiac cycle.

The second approach is to acquire gated imaging data with a minimum of two echoes per repetition time (TR) in order to derive an image basically reflecting T2* effects, thus reducing signal variations due to T1 saturation effects (Beissner et al., 2010; Beissner et al., 2011; Zhang et al., 2006). The technique is based on the calculation of the apparent T2* using a mathematical approach, or by computing the quotient of every two images per repetition time. In any case, according to Breissner et al. (2011), pre-processing for the smoothing of data and optimal normalization is necessary.

8.3.1.2 Acquisition-Based Image Corrections

Physiologically-induced image artifacts can be mitigated using modifications of the pulse sequence parameters of BOLD imaging, although the major difficulty that arises, is to maintain an adequately high SNR. As mentioned earlier, the EPI acquisitions that are typically used for fMRI are vulnerable to phase errors from field variations due to motion and hence multiple variations of the T2*-weighted sequences have been proposed to resolve these issues, such as multi-echo EPI and parallel imaging acquisitions using slice-dependent TE sequences. Especially at higher magnetic fields (>3 T) these can be easily implemented as modified versions of Gradient Echo sequences, using multi-channel coils (Domsch et al., 2013; Zaca et al., 2014).

The multi-channel array coils with parallel imaging may increase the contrast to noise ratio of the EPI data using reduced echo train lengths to acquire additional echoes for the same TR. It has been shown that multi-echo imaging techniques can minimize physiological signal fluctuation in and around the brainstem (Kundu et al., 2012).

Another approach to minimizing the influences of geometric distortions is the use of the Point Spread Function (PSF) (Zaitsev et al., 2004). Additional phase encoding gradient acquisitions in the x, y, and z directions are used to map all three-dimensional PSFs of each voxel. Then PSFs are used to describe each voxel's distribution of intensities and hence spatial information, which allows the geometrical distortion to be corrected for. The downside is that this technique obviously requires additional scan and processing times, limiting its use in clinical routine.

The last popular imaging technique to mitigate these influences is the use of the so-called "navigator echoes" (Pfeuffer et al., 2002). The general idea is to acquire an additional non-phase-encoded navigator echo before the image read-out in order to adjust the phase shift of the data produced by the motion-induced time-varying magnetic field. The comparison of the navigator echo to the central line of the main k-space data then be used to correct the phase of all other echoes recorded and yield enough information to remove the geometric distortion in the phase-encoding direction. Nevertheless, caution is needed since navigator echoes are not so easy to manipulate.

8.3.1.3 Calibration

8.3.1.3.1 Calibration Scans

The idea behind calibration scans is to record and identify physiological signals in the absence of any external brain stimulation and then remove them from the task fMRI results during the data post-processing. The procedure is sometimes called a "clean-up" since it utilizes the physiological recordings to remove the estimates from each voxel's time series using linear regression, thus "cleaning up" the fMRI data (Murphy et al., 2013). This is done using a linear regression procedure to model the BOLD signal as a time series, and to fit it to the "influenced" data. This fit is then subtracted from the original data to remove the associated physiological variance.

While this approach increases the effectiveness of fMRI experiments, there are some associated disadvantages, such as (1) the extra time needed to acquire the additional scan for the calibration, and more importantly (2) the possibility of removing signals that are recognized as physiological noise but may be indeed signals correlated to the induced brain function.

The same goal of physiological noise sources removal can be mitigated by several approaches and in different data levels. Hu et al. (1995) proposed a method to retrospectively estimate and correct the primary fluctuation effects of cardiac and respiratory cycles using pulse oximeters and respiratory bellows in the k-space (called RETROKCOR), while Glover et al. (2000) proposed the same approach but in the image space data and argued that for standard acquisitions it worked better than the k-space approach (called RETROCOIR). This is most probably due to the fact that corrections of single k-space points affect all voxels, potentially introducing spatially correlated noise in all data (Murphy et al., 2013).

Both the RETROKCOR from Hu et al., and the RETROCOIR from Glover et al., obviously depend on the model used to approximate the physiological fluctuations effects. These signal fluctuations can alternatively be reduced by the use of a model-free adaptive filtering, provided they are repetitive and their timing is known (Deckers et al., 2006). The performance of this approach has been proposed to be at least equivalent to the RETROCOIR method, under the assumption that "each occurrence of the quasi-periodic disturbance ('event') leads to an artifact with spatial and temporal signal perturbation characteristics that only depend on the timing of the event relative to MRI data acquisition" (Deckers et al., 2006).

There are also a number of pre-processing strategies used for reducing physiological noise implemented after the acquisition, including (1) *temporal filtering* (theoretically, it is possible to filter and remove signals with certain frequencies, such as the cardiac and respiratory cycle) and (2) *denoising with independent component analysis (ICA)* (theoretically, using machine learning algorithms and automated classification methods, any dataset can be "cleaned up" or decomposed into its constituent sources). Nevertheless, a potential limitation regarding temporal filtering is that the acquisition rate of fMRI data does not adequately sample the cardiac and respiratory variations that might be aliased into the BOLD signal frequency (Zaca et al., 2014). Hence, removing physiological noise will also remove portions of the BOLD processes, decreasing the overall signal. To this effect, special attention must be given in the subsequent statistics when denoising with ICA in order to avoid falsely inflated statistics (Brooks et al., 2013).

Eventually the question that arises is: which one should be used?

Unfortunately, there is not an easy answer. It depends on various factors, such as the subject group, the available software and scanner, as well as the specific fMRI experiment.

Overall there are several different "ready-to-use" software packages for physiological noise mitigation of fMRI data, either as specific tools embedded in other software, or as stand-alone packages. As an indication these are illustrated (but not limited to) in the following updated table (Table 8.1) including data from Brooks et al. (2013). For the interested reader, the review paper by Brooks et al. (2013) provides extended information and details for physiological noise in brainstem fMRI.

8.4 Interpretation Limitations

Probably the major concern raised with respect to the interpretation of BOLD fMRI is the uncertainty of neural activity localization. The fundamental goal in fMRI is to be able to precisely detect transient hemodynamic changes involving a number of mechanisms following neuronal activation, such as changes in blood flow, volume, and oxygenation. The uncertainty

TABLE 8.1 A Selection of Software Packages for Physiological Noise Mitigation and Processing of fMRI Data

TOOLS	URL
Stand-Alone Packages	
RETROICOR tool	http://www.mccauslandcenter.sc.edu/crnl/tools/part
PART (Physiological Artifact Removal Tool)	http://journals.plos.org/plosone/article?id=10.1371/journal.pone.0001751#pone.0001751.s001
PhysioNoise	https://github.com/timothyv/Physiological-Log-Extraction-for-Modeling--PhLEM--Toolbox
PhLEM (Physiological Log Extraction for Modeling)	https://www.tnu.ethz.ch/de/software/tapas.html
TAPAS PhysIO Toolbox	http://www.cs.tut.fi/~jupeto/software.html
ICA Artifact Remover NIAK (NeuroImaging Analysis Kit)	http://www.mrijournal.com/article/S0730-725X(06)00342-0/abstract
CORSICA (CORrection of Structured noise using spatial Independent Component Analysis)	http://www.nitrc.org/projects/pestica
PESTICA (Physiologic EStimation by Temporal ICA)	http://cbi.nyu.edu/software/
Specific Tools	
RETROICOR (RETROspective Image CORrection) (AFNI)	http://afni.nimh.nih.gov/pub/dist/doc/program_help/3dretroicor.html
RETROICOR (FreeSurfer)	https://github.com/neurodebian/freesurfer/blob/master/fsfast/toolbox/fast_retroicor.m
DRIFTER (Dynamic RetrospectIve FilTERing) (SPM ToolBox)	http://becs.aalto.fi/en/research/bayes/drifter/
PNM (physiological noise model) (Part of FSL)	http://fsl.fmrib.ox.ac.uk/fsl/fslwiki/PNM
FIX (FMRIB's ICA-based X-noisifier) (Part of FSL)	http://fsl.fmrib.ox.ac.uk/fsl/fslwiki/FIX
Processing Software	
AFNI (Analysis of Functional NeuroImages)	https://afni.nimh.nih.gov/
BrainVoyager (analysis and visualization of structural and functional magnetic resonance imaging data)	http://www.brainvoyager.com/
FMRISTAT (A general statistical analysis for fMRI data) (From McGill University)	http://www.math.mcgill.ca/keith/fmristat/
BIC (Brain Imaging Centre) Software	https://www.mcgill.ca/bic/software
Brain Connectivity Toolbox (Most widely used to analyze resting-state fMRI studies)	https://sites.google.com/site/bctnet/
BrainSuite (A collection of open source software tools)	http://brainsuite.org/
BrainVISA (A free neuroimaging software platform for mass data analysis) (IFR-France)	http://www.brainvoyager.com/
FreeSurfer is an open source software suite for processing and analyzing brain MR images	http://www.brainvoyager.com/
FSL (FMRIB Software Library)	https://fsl.fmrib.ox.ac.uk/fsl/fslwiki/
SPM (Statistical Parametric Mapping)	http://www.fil.ion.ucl.ac.uk/spm/
REST (RESting-state fMRI data analysis Toolkit) is a group of applications based on MATLAB and SPM8 for evaluation of resting-state fMRI data	http://restfmri.net/forum/index.php?q=rest
MIALAB (Medical Image Analysis Lab)	http://mialab.mrn.org/

arises from the fact that the extent of the fMRI hemodynamic response around the active sites, at submillimeter resolution, remains controversial (Duong et al., 2000; Zaca et al., 2014).

```
Why?
```

The explanation is that these hemodynamic changes can propagate from the capillary beds adjacent to the site of neuronal activation beyond the actual activated sites, hence the BOLD response may appear diffused far from the active sites (Duong et al., 2000; Duong et al., 2001).

In fact, we need to realize that the BOLD contrast arises from a complex interplay between cerebral blood flow, cerebral blood volume, and cerebral metabolic rate of oxygen consumption on a spatial scale that lumps together hundreds of thousands of neurons in each MR imaging voxel (Zaca et al., 2014). In that sense, it is very much dependent on the local vasculature and its spatial characteristics (Kim and Ogawa, 2012; Moon et al., 2013), and obviously depends on the area of activation. More specifically, the spatial specificity can be decreased because of the so-called venous effects. Especially in areas of high vascular density, these effects increase the probability of false-positive activation since the strongest BOLD signal change is generated near draining veins (Bianciardi et al., 2011). Nevertheless, at the increasingly high magnetic fields used clinically, spatial resolution can be improved using spin-echo sequences (Kim and Ogawa, 2012). Especially with the introduction of ultrahigh field magnets (>7 T), susceptibility weighted imaging has been used to remove the venous signal, in terms of vascular activation masking. Then, the venous contribution is expected to decrease since, with higher magnetic fields and longer TEs, the extravascular contribution increases and the intravascular contribution decreases; therefore, the spatial specificity to parenchyma improves (Kim and Ogawa, 2012). Another approach to increasing the spatial specificity is to enhance the statistical significance threshold. However, this might cause the opposite error of false-negative activation since the background signal might wash out and should be avoided.

The extent of activation foci is limited by another difficulty, which relates to the high inter- and intra-subject variability observed with fMRI, even with constant scanning parameters and consistent task performance (Goodyear et al., 2014). Especially regarding fMRI pre-surgical mapping, perhaps the most serious concern is the lack of systematic establishment and verification of the technique in different patient types and using specific activation paradigms.

Again, the extent of activation depends on a statistical threshold, which in turn is determined by the quality of the BOLD signal and the accomplished SNR. This is called activation map thresholding, and it has been shown that although the spatial distribution of BOLD t-value statistical activation maps can be highly variable across subjects and scan duration, the use of relative activation maps, (e.g., the use of 40% of the most active voxels) resulted in highly reproducible results both within individual subjects and across different subjects (Voyvodic et al., 2009).

8.5 Quality Assurance in fMRI

All MRI studies should be performed in a well-maintained scanner and a quality assurance (QA) program should be mandatory to obtain optimal images in any clinical environment. fMRI studies are particularly demanding and should be given extra attention since they utilize fast imaging methods such as echo-planar imaging (EPI) or spiral acquisitions that usually push the scanner hardware to its limits. Hence, besides good experimental design, correct technical execution of the scan and minimal subject and physiological motion, the fMRI experiment should include a well-designed QA program to ensure optimal data extraction.

There are several specifically designed QA programs for fMRI studies and the interested reader should refer to them for an analytical description and guidance. (Friedman and Glover, 2006; Greve et al., 2011; Olsrud et al., 2008; Stöcker et al., 2005; Sutton et al., 2008). Nevertheless, it is important that each research site establish its own QA performance criteria and the afore-mentioned QA studies may be useful in guiding the establishment of such criteria. Ideally, fMRI QA scans should be run daily and a good recommended practice is to have the QA scans acquired at the beginning of each day prior to any scheduled scans.

8.6 Conclusion

Functional MRI has made a huge clinical impact during the last decade for the noninvasive evaluation of brain functionality and connectivity, and it is safe to argue that in the following years it may further improve the effectiveness of human brain studies in an accelerated fashion.

Nevertheless, there should be an even greater degree of attention to the data analysis methods and the new acquisition protocols used since, undoubtedly, there are many issues that need to be resolved before clinical fMRI can be fully implemented in the clinical routine. Although the vast majority of fMRI studies are carefully conducted, the correct and consistent application of the analysis pipeline can often be very challenging. All fMRI studies will continuously strike a delicate balance between over-analyzing or underestimating the results, as well as put forth a continuous effort to recognize and cope with technical and physiological artifacts and sources of error.

References

Bandettini, P. A. (2009). Functional MRI limitations and aspirations. In E. Pöppel, B. Gulyás, E. Kraft, A. Müller, and E. Pöppel (Eds.), *Neural Correlates of Thinking*. Berlin, Heidelberg: Springer.

Bandettini, P. A. and Wong, E. C. (1997). A hypercapnia-based normalization method for improved spatial localization of human brain activation with fMRI. *NMR in Biomedicine*, 10(4–5), 197–203. doi:10.1002/(sici)1099-1492(199706/08)10:4/5<197::aid-nbm466>3.0.co;2-s

Beissner, F., Baudrexel, S., Volz, S., and Deichmann, R. (2010). Dual-echo EPI for non-equilibrium fMRI— Implications of different echo combinations and masking procedures. *NeuroImage*, 52(2), 524–531. doi:10.1016/j.neuroimage.2010.04.243

Beissner, F., Deichmann, R., and Baudrexel, S. (2011). FMRI of the brainstem using dual-echo EPI. *NeuroImage*, 55(4), 1593–1599. doi:10.1016/j.neuroimage.2011.01.042

Bellgowan, P. S., Saad, Z. S., and Bandettini, P. A. (2003). Understanding neural system dynamics through task modulation and measurement of functional MRI amplitude, latency, and width. *Proceedings of the National Academy of Sciences*, 100(3), 1415–1419. doi:10.1073/pnas.0337747100

Bianciardi, M., Fukunaga, M., Gelderen, P. V., Zwart, J. A., and Duyn, J. H. (2011). Negative BOLD-fMRI signals in large cerebral veins. *Journal of Cerebral Blood Flow & Metabolism*, 31(2), 401–412. doi:10.1038/jcbfm.2010.164

Birn, R. M., Diamond, J. B., Smith, M. A., and Bandettini, P. A. (2006). Separating respiratory-variation-related fluctuations from neuronal-activity-related fluctuations in fMRI. *NeuroImage*, 31(4), 1536–1548. doi:10.1016/j.neuroimage.2006.02.048

Brooks, J. C., Faull, O. K., Pattinson, K. T., and Jenkinson, M. (2013). Physiological noise in brainstem fMRI. *Frontiers in Human Neuroscience*, 7, 623. doi:10.3389/fnhum.2013.00623

Brosch, J., Talavage, T., Ulmer, J., and Nyenhuis, J. (2002). Simulation of human respiration in fMRI with a mechanical model. *IEEE Transactions on Biomedical Engineering*, 49(7), 700–707. doi:10.1109/tbme.2002.1010854

Buxton, R. B., Uludağ, K., Dubowitz, D. J., and Liu, T. T. (2004). Modeling the hemodynamic response to brain activation. *NeuroImage, 23,* Suppl 1, S220–233. doi:10.1016/j.neuroimage.2004.07.013

Chiacchiaretta, P. and Ferretti, A. (2015). Resting state BOLD functional connectivity at 3T: Spin echo versus gradient echo EPI. *PLOS One, 10*(3), e0120398. doi:10.1371/journal.pone.0120398

Chiarelli, P. A., Bulte, D. P., Wise, R., Gallichan, D., and Jezzard, P. (2007). A calibration method for quantitative BOLD fMRI based on hyperoxia. *NeuroImage, 37*(3), 808–820. doi:10.1016/j.neuroimage.2007.05.033

Dagli, M. S., Ingeholm, J. E., and Haxby, J. V. (1999). Localization of cardiac-induced signal change in fMRI. *NeuroImage, 9*(4), 407–415. doi:10.1006/nimg.1998.0424

Deckers, R. H., Gelderen, P. V., Ries, M., Barret, O., Duyn, J. H., Ikonomidou, V. N., and Zwart, J. A. (2006). An adaptive filter for suppression of cardiac and respiratory noise in MRI time series data. *NeuroImage, 33*(4), 1072–1081. doi:10.1016/j.neuroimage.2006.08.006

Domsch, S., Linke, J., Heiler, P. M., Kroll, A., Flor, H., Wessa, M., and Schad, L. R. (2013). *Increased BOLD sensitivity in the orbitofrontal cortex using slice-dependent echo times at 3T. Magnetic Resonance Imaging, 31*(2), 201–211. doi:10.1016/j.mri.2012.06.020

Duong, T. Q., Kim, D., Uğurbil, K., and Kim, S. (2000). Spatiotemporal dynamics of the BOLD fMRI signals: Toward mapping submillimeter cortical columns using the early negative response. *Magnetic Resonance in Medicine, 44*(2), 231–242. doi:10.1002/1522-2594(200008)44:2<231::aid-mrm10>3.3.co;2-k

Duong, T. Q., Kim, D., Uğurbil, K., and Kim, S. (2001). Localized cerebral blood flow response at submillimeter columnar resolution. *Proceedings of the National Academy of Sciences, 98*(19), 10904–10909. doi:10.1073/pnas.191101098

Engel, S. (1997). Retinotopic organization in human visual cortex and the spatial precision of functional MRI. *Cerebral Cortex, 7*(2), 181–192. doi:10.1093/cercor/7.2.181

Friedman, L. and Glover, G. H. (2006). Report on a multicenter fMRI quality assurance protocol. *Journal of Magnetic Resonance Imaging, 23*(6), 827–839. doi:10.1002/jmri.20583

Friston, K. J., Williams, S., Howard, R., Frackowiak, R. S., and Turner, R. (1996). Movement-related effects in fMRI time-series. *Magnetic Resonance in Medicine, 35*(3), 346–355. doi:10.1002/mrm.1910350312

Glover, G. H., Li, T., and Ress, D. (2000). Image-based method for retrospective correction of physiological motion effects in fMRI: RETROICOR. *Magnetic Resonance in Medicine, 44*(1), 162–167. doi:10.1002/1522-2594(200007)44:1<162::aid-mrm23>3.3.co;2-5

Gonzales, M. M., Tarumi, T., Mumford, J. A., Ellis, R. C., Hungate, J. R., Pyron, M., and Haley, A. P. (2014). Greater BOLD response to working memory in endurance-trained adults revealed by breath-hold calibration. *Human Brain Mapping, 35*(7), 2898–2910. doi:10.1002/hbm.22372

Goodyear, B., Liebenthal, E., and Mosher, V. (2014). Active and passive fMRI for presurgical mapping of motor and language cortex. In T. D. Papageorgiou, G. I. Christopoulos, and S. M. Smirnakis (Eds.), *Advanced Brain Neuroimaging Topics in Health and Disease—Methods and Applications.* doi:10.5772/58269

Greve, D. N., Mueller, B. A., Liu, T., Turner, J. A., Voyvodic, J., Yetter, E. et al. (2011). A novel method for quantifying scanner instability in fMRI. *Magnetic Resonance in Medicine, 65*(4), 1053–1061. doi:10.1002/mrm.22691

Halai, A. D., Welbourne, S. R., Embleton, K., and Parkes, L. M. (2014). A comparison of dual gradient-echo and spin-echo fMRI of the inferior temporal lobe. *Human Brain Mapping, 35*(8), 4118–4128. doi:10.1002/hbm.22463

Handwerker, D. A., Gazzaley, A., Inglis, B. A., and Desposito, M. (2007). Reducing vascular variability of fMRI data across aging populations using a breathholding task. *Human Brain Mapping, 28*(9), 846–859. doi:10.1002/hbm.20307

Hu, X., Le, T. H., Parrish, T., and Erhard, P. (1995). Retrospective estimation and correction of physiological fluctuation in functional MRI. *Magnetic Resonance in Medicine, 34*(2), 201–212. doi:10.1002/mrm.1910340211

Hutton, C., Bork, A., Josephs, O., Deichmann, R., Ashburner, J., and Turner, R. (2002). Image distortion correction in fMRI: A quantitative evaluation. *NeuroImage, 16*(1), 217–240. doi:10.1006/nimg.2001.1054

Hutton, C., Josephs, O., Stadler, J., Featherstone, E., Reid, A., Speck, O., and Weiskopf, N. (2011). The impact of physiological noise correction on fMRI at 7T. *NeuroImage*, *57*(1), 101–112. doi:10.1016/j.neuroimage.2011.04.018

In, M. and Speck, O. (2012). Highly accelerated PSF-mapping for EPI distortion correction with improved fidelity. *MAGMA*, *25*(3), 183–192.

Jezzard, P. (1999). Physiological noise: Strategies for correction. In C. Moonen, P. A. Bandettini (Eds.), *Functional MRI* (pp. 171–179). Heidelberg, Germany: Springer-Verlag.

Jezzard, P. (2012). Correction of geometric distortion in fMRI data. *NeuroImage*, *62*(2), 648–651. doi:10.1016/j.neuroimage.2011.09.010

Jezzard, P. and Balaban, R. S. (1995). Correction for geometric distortion in echo planar images from B0 field variations. *Magnetic Resonance in Medicine*, *34*(1), 65–73. doi:10.1002/mrm.1910340111

Jochimsen, T. H., Norris, D. G., Mildner, T., and Möller, H. E. (2004). *Quantifying the intra- and extra-vascular contributions to spin-echo fMRI at 3 T. Magnetic Resonance in Medicine*, *52*(4), 724–732. doi:10.1002/mrm.20221

Kim, S. and Ogawa, S. (2012). Biophysical and physiological origins of blood oxygenation level-dependent fMRI signals. *Journal of Cerebral Blood Flow and Metabolism*, *32*(7), 1188–1206. doi:10.1038/jcbfm.2012.23

Krüger, G. and Glover, G. H. (2001). Physiological noise in oxygenation-sensitive magnetic resonance imaging. *Magnetic Resonance in Medicine*, *46*(4), 631–637. doi:10.1002/mrm.1240.abs

Kundu, P., Inati, S. J., Evans, J. W., Luh, W., and Bandettini, P. A. (2012). Differentiating BOLD and non-BOLD signals in fMRI time series using multi-echo EPI. *NeuroImage*, *60*(3), 1759–1770. doi:10.1016/j.neuroimage.2011.12.028

Li, X., Lu, Z., Tjan, B. S., Dosher, B. A., and Chu, W. (2008). Blood oxygenation level-dependent contrast response functions identify mechanisms of covert attention in early visual areas. *Proceedings of the National Academy of Sciences*, *105*(16), 6202–6207. doi:10.1073/pnas.0801390105

Malinen, S., Schürmann, M., Hlushchuk, Y., Forss, N., and Hari, R. (2006). Improved differentiation of tactile activations in human secondary somatosensory cortex and thalamus using cardiac-triggered fMRI. *Experimental Brain Research*, *174*(2), 297–303. doi:10.1007/s00221-006-0465-z

Mikl, M., Mareček, R., Hluštík, P., Pavlicová, M., Drastich, A., Chlebus, P. et al. (2008). Effects of spatial smoothing on fMRI group inferences. *Magnetic Resonance Imaging*, *26*(4), 490–503. doi:10.1016/j.mri.2007.08.006

Moon, C. H., Fukuda, M., and Kim, S. (2013). Spatiotemporal characteristics and vascular sources of neural-specific and -nonspecific fMRI signals at submillimeter columnar resolution. *NeuroImage*, *64*, 91–103. doi:10.1016/j.neuroimage.2012.08.064

Murphy, K., Birn, R. M., and Bandettini, P. A. (2013). Resting-state fMRI confounds and cleanup. *NeuroImage*, *80*, 349–359. doi:10.1016/j.neuroimage.2013.04.001

Murphy, K., Bodurka, J., and Bandettini, P. A. (2007). How long to scan? The relationship between fMRI temporal signal to noise ratio and necessary scan duration. *NeuroImage*, *34*(2), 565–574. doi:10.1016/j.neuroimage.2006.09.032

Mutch, W. A., Ellis, M. J., Graham, M. R., Wourms, V., Raban, R., Fisher, J. A., and Ryner, L. (2014). Brain MRI CO2 stress testing: A pilot study in patients with concussion. *PLOS One*, *9*(7), e102181. doi:10.1371/journal.pone.0102181

Norris, D. G. (2006). Principles of magnetic resonance assessment of brain function. *Journal of Magnetic Resonance Imaging*, *23*(6), 794–807. doi:10.1002/jmri.20587

Norris, D. G. (2012). Spin-echo fMRI: The poor relation? *NeuroImage*, *62*(2), 1109–1115. doi:10.1016/j.neuroimage.2012.01.003

Ojemann, J. G., Akbudak, E., Snyder, A. Z., Mckinstry, R. C., Raichle, M. E., and Conturo, T. E. (1997). Anatomic localization and quantitative analysis of gradient refocused echo-planar fMRI susceptibility artifacts. *NeuroImage*, *6*(3), 156–167. doi:10.1006/nimg.1997.0289

Olman, C. A., Davachi, L., and Inati, S. (2009). Distortion and signal loss in medial temporal lobe. *PLOS One*, *4*(12), e8160. doi:10.1371/journal.pone.0008160

Olsrud, J., Nilsson, A., Mannfolk, P., Waites, A., and Ståhlberg, F. (2008). A two-compartment gel phantom for optimization and quality assurance in clinical BOLD fMRI. *Magnetic Resonance Imaging*, 26(2), 279–286. doi:10.1016/j.mri.2007.06.010

Parkes, L. M., Schwarzbach, J. V., Bouts, A. A., Deckers, R. H., Pullens, P., Kerskens, C. M., and Norris, D. G. (2005). Quantifying the spatial resolution of the gradient echo and spin echo BOLD response at 3 Tesla. *Magnetic Resonance in Medicine*, 54(6), 1465–1472. doi:10.1002/mrm.20712

Peck, K. K., Bradbury, M., Petrovich, N., Hou, B. L., Ishill, N., Brennan, C., and Holodny, A. I. (2009). Presurgical evaluation of language using functional magnetic resonance imaging in brain tumor patients with previous surgery. *Neurosurgery*, 64(4), 644–653. doi:10.1227/01.neu.0000339122.01957.0a

Pfeuffer, J., Moortele, P. V., Ugurbil, K., Hu, X., and Glover, G. H. (2002). Correction of physiologically induced global off-resonance effects in dynamic echo-planar and spiral functional imaging. *Magnetic Resonance in Medicine*, 47(2), 344–353. doi:10.1002/mrm.10065

Raj, D., Anderson, A. W., and Gore, J. C. (2001). Respiratory effects in human functional magnetic resonance imaging due to bulk susceptibility changes. *Physics in Medicine and Biology*, 46(12), 3331–3340. doi:10.1088/0031-9155/46/12/318

Schwarzbauer, C., Mildner, T., Heinke, W., Brett, M., and Deichmann, R. (2010). Dual echo EPI—The method of choice for fMRI in the presence of magnetic field inhomogeneities? *NeuroImage*, 49(1), 316–326. doi:10.1016/j.neuroimage.2009.08.032

Shmuel, A. and Maier, A. (2015). Locally measured neuronal correlates of functional MRI signals. In K. Ugurbil, L. Berliner, and K. Uludag (Eds.), *FMRI: From Nuclear Spins to Brain Functions*. New York, NY: Springer.

Shmuel, A., Yacoub, E., Chaimow, D., Logothetis, N. K., and Ugurbil, K. (2007). Spatio-temporal point-spread function of fMRI signal in human gray matter at 7 Tesla. *NeuroImage*, 35(2), 539–552. doi:10.1016/j.neuroimage.2006.12.030

Smith, S. M., Beckmann, C. F., Khorshidi, G. S., Miller, K. L., Nichols, T. E., Ramsey, J. et al. (2011). Network modelling methods for FMRI. *NeuroImage*, 54(2), 875–891.

Soellinger, M., Ryf, S., Boesiger, P., and Kozerke, S. (2007). Assessment of human brain motion using CSPAMM. *Journal of Magnetic Resonance Imaging*, 25(4), 709–714. doi:10.1002/jmri.20882

Stöcker, T., Schneider, F., Klein, M., Habel, U., Kellermann, T., Zilles, K., and Shah, N. J. (2005). Automated quality assurance routines for fMRI data applied to a multicenter study. *Human Brain Mapping*, 25(2), 237–246. doi:10.1002/hbm.20096

Sutton, B. P., Goh, J., Hebrank, A., Welsh, R. C., Chee, M. W., and Park, D. C. (2008). Investigation and validation of intersite fMRI studies using the same imaging hardware. *Journal of Magnetic Resonance Imaging*, 28(1), 21–28. doi:10.1002/jmri.21419

Thomason, M. E., Burrows, B. E., Gabrieli, J. D., and Glover, G. H. (2005). Breath holding reveals differences in fMRI BOLD signal in children and adults. *NeuroImage*, 25(3), 824–837. doi:10.1016/j.neuroimage.2004.12.026

Thomason, M. E., Foland, L. C., and Glover, G. H. (2007). Calibration of BOLD fMRI using breath holding reduces group variance during a cognitive task. *Human Brain Mapping*, 28(1), 59–68.

Thomason, M. E. and Glover, G. H. (2008). Controlled inspiration depth reduces variance in breath-holding-induced BOLD signal. *NeuroImage*, 39(1), 206–214.

Thompson, S. K., Engel, S. A., and Olman, C. A. (2014). Larger neural responses produce BOLD signals that begin earlier in time. *Frontiers in Neuroscience*, 8, 159. doi:10.3389/fnins.2014.00159

Triantafyllou, C., Hoge, R., Krueger, G., Wiggins, C., Potthast, A., Wiggins, G., and Wald, L. (2005). Comparison of physiological noise at 1.5 T, 3 T and 7 T and optimization of fMRI acquisition parameters. *NeuroImage*, 26(1), 243–250. doi:10.1016/j.neuroimage.2005.01.007

Voyvodic, J. T., Petrella, J. R., and Friedman, A. H. (2009). FMRI activation mapping as a percentage of local excitation: Consistent presurgical motor maps without threshold adjustment. *Journal of Magnetic Resonance Imaging*, 29(4), 751–759. doi:10.1002/jmri.21716

Welvaert, M. and Rosseel, Y. (2013). On the definition of signal-to-noise ratio and contrast-to-noise ratio for fMRI data. *PLOS One*, 8(11), e77089. doi:10.1371/journal.pone.0077089

Wise, R. G., Ide, K., Poulin, M. J., and Tracey, I. (2004). Resting fluctuations in arterial carbon dioxide induce significant low frequency variations in BOLD signal. *NeuroImage, 21*(4), 1652–1664. doi:10.1016/j.neuroimage.2003.11.025

Zaca, D., Agarwal, S., Gujar, S. K., Sair, H. I., and Pillai, J. J. (2014). Special considerations/technical limitations of blood-oxygen-level-dependent functional magnetic resonance imaging. *Neuroimaging Clinics of North America, 24*(4), 705–715. doi:10.1016/j.nic.2014.07.006

Zaitsev, M., Hennig, J., and Speck, O. (2004). Point spread function mapping with parallel imaging techniques and high acceleration factors: Fast, robust, and flexible method for echo-planar imaging distortion correction. *Magnetic Resonance in Medicine, 52*(5), 1156–1166. doi:10.1002/mrm.20261

Zhang, W., Mainero, C., Kumar, A., Wiggins, C. J., Benner, T., Purdon, P. L., and Sorensen, A. G. (2006). Strategies for improving the detection of fMRI activation in trigeminal pathways with cardiac gating. *NeuroImage, 31*(4), 1506–1512. doi:10.1016/j.neuroimage.2006.02.033

9

The Role of Multiparametric MR Imaging—Advanced MR Techniques in the Assessment of Cerebral Tumors

9.1 Introduction

Conventional MR imaging findings in a number of intracranial masses are sometimes non-specific and, despite the recent technological advancements, tumor characterization and grading can be a challenging process. Continual efforts are being made to assess the utility of advanced imaging techniques and enable the extraction of quantifiable features for the assessment of cerebral tumors in a reliable manner with consistent correlation and specific endpoints.

The parameters extracted from the advanced techniques discussed in chapters 1, 3, 5, and 7 provide significant structural and functional information in a microscopic and cellular level, highlighting aspects of the underlying brain pathophysiology; however, their accurate interpretation is not always straightforward as similarities may exist between pathologies, and one should be very careful in correctly combining and evaluating all the available MR data. When found to have significant correlation, these parameters may possibly serve as endpoints for the assessment of the severity, degree of change, or status of a brain disease, relative to normal.

Unfortunately, none of the techniques alone has proven to be robust enough to be solely used for the aforementioned purposes.

At this point, it should be realized that over the years, MR systems have evolved from imaging modalities to advanced computational systems producing a variety of numerical parameters, which can be difficult for the radiologist to accurately evaluate during the clinical routine.

Hence, despite the indisputable contribution of advanced techniques to the preoperative assessment of cerebral pathology (Calli et al., 2006; Chang et al., 2002; De Belder et al., 2012; Liu et al., 2011; Svolos et al., 2013), the reported results in the literature can be conflicting and may further complicate clinical decision making (Gillard et al., 2005; Lu et al., 2004; Server et al., 2009; Tsougos et al., 2012; Wang et al., 2011). It is precisely these controversies that reflect the complex underlying pathophysiologic mechanisms that are present in cerebral lesions and may prevent the clear discrimination between pathologies.

Usually, the most critical elements in the determination of tumor grade and prognosis are tumor cellularity and vascularity. These elements can be quantified using diffusion and perfusion techniques; however, as they are closely correlated, their evaluation and interpretation is difficult on the basis of individual numeric parameters. Combining the above techniques with magnetic resonance spectroscopy into a multiparametric evaluation, including metabolic characteristics, is expected to substantially increase diagnostic accuracy.

Data analysis using conventional methods such as statistical significances correlations of the related parameters between different tumor groups may be efficient in some cases. However, in more demanding diagnostic problems such as pathologies mimicking tumors or lesions with identical pathophysiological profiles, their efficiency might be limited. It is interesting to note that recently, research and clinical interest has been focused on the incremental diagnostic and predictive value of a multi-parametric approach using advanced MRI techniques. Different methods of data analysis have been evaluated, such as logistic regression (LR) and receiver operating characteristic (ROC) analysis (Tsougos et al., 2012; Zonari et al., 2007; Svolos et al., 2014), as well as more sophisticated techniques like machine learning algorithms (Devos et al., 2005; Georgiadis et al., 2011; Law et al., 2003; Zacharaki et al., 2011), using various parametric combinations.

The combination of DWI and Magnetic Resonance Spectroscopic Imaging (MRSI) has been reported to increase the accuracy of preoperative differentiation of low-grade gliomas (LGG) versus high-grade gliomas (HGG) (Server et al., 2011). In this study investigators used a model consisting of four factors, namely the intratumoral mean ADC and maximum ADC and peritumoral Cho/Cr and Cho/NAA ratios, and resulted in 92.5% accuracy, 91.5% sensitivity, 100% specificity and positive predictive value (PPV) and 60% negative predictive value (NPV). Moreover, Wang et al. (2011) investigated the differentiation of glioblastomas, solitary brain metastases, and primary cerebral lymphomas (PCLs). The authors showed that the best model to discriminate glioblastomas from other lesions consisted of ADC and FA from the enhancing region of the tumors. The accuracy, sensitivity, and specificity scored were 93.8%, 89%, and 93%, respectively. Additionally, the best model to differentiate PCLs from metastases consisted of ADC from the enhancing regions and the planar anisotropy coefficient from the immediate peritumoral area. The accuracy, sensitivity and specificity were 90.9%, 77%, and 94%, respectively. Similarly, Zonari et al. (2007) showed that the differentiation of LGG and HGG is more efficient if DWI, DSCI, and MRSI data are combined than evaluated independently.

Therefore, it seems that multi-parametric analysis may substantially improve diagnostic accuracies over conventional MRI alone, and highlight the underlying pathophysiology. However, as mentioned before, this process is quite demanding and time-consuming due to the numeric nature of the acquired MR data. Recent studies have reported that machine learning techniques may be used as an automated computer analysis tool in order to aid tumor diagnosis (Devos et al., 2005; Hu et al., 2011; Law et al., 2003; Tsolaki et al., 2013). The use of such techniques allows the manipulation and evaluation of a large amount of quantitative data during clinical practice. A variety of features, such as morphological (e.g., tumor shape and texture) and conventional (e.g., signal intensity), extracted from different MR sequences have been evaluated with very interesting results (Vamvakas et al., 2016; Verma et al., 2008; Zacharaki et al., 2009). However, the most important aspect of machine learning techniques is their additional ability to provide predictive outcomes in contrast to conventional statistical methods, which are limited to producing diagnostic results retrospectively.

In conclusion, brain tumors constitute a diverse group of lesions originating from different cells within the central nervous system (CNS) and include a variety of histologic types with markedly different tumor growth rates. Hence, a multiparametric approach utilizing all the advanced MR techniques to evaluate the different underlying pathophysiological characteristics seems promising. In this chapter, we will discuss the main types of brain tumors in an attempt to illustrate the potential ability of DWI/DTI, MRS, perfusion and fMRI techniques to contribute to a more accurate diagnosis.

9.2 Gliomas

Focus Point

DWI

- Ambiguous results in the differentiation of lower and higher glioma grades.
- Large variations of ADC values exist between the two groups.

DTI

- FA can distinguish HGG from LGG.
- High anisotropy implies symmetrically organized tissue.

PWI

- rCBV provides robust differentiation between LGG and HGG based on microvasculature differences.
- Progressive increase in rCBV from lower to higher glioma grades.
- Overlapping in rCBV values among intermediate grades might be observed.

MRS

- Differentiation of LGG vs HGG, using maximal Cho/Cr, Cho/NAA and minimum NAA/Cr.
- Higher levels of Myo-inositol in LGG vs. HGG.
- MRS specificity still relatively low.

fMRI

- Localization of brain function for pre-surgical planning.

Gliomas represent the most common primary cerebral neoplasms and their preoperative assessment and grading is of outmost importance for therapeutic decision making. "Glioma" is a general term used to describe any tumor that arises from the supportive tissue of the brain called "glia," which comes from the Greek word ("γλία") or "gluey," meaning that it serves as the connective substance to keep the neurons in place and well-functioning. Hence gliomas may arise from the supporting glial cells in the brain and their predominant cell type determines their pathological classification. There are two major categories of gliomas:

1. Low-grade gliomas are well-differentiated (not anaplastic); these are benign and portend a better prognosis for the patient.
2. High-grade gliomas are undifferentiated or anaplastic; these are malignant and carry a worse prognosis for the patient.

Low-grade gliomas (LGG) consist of Grade I, which progress very slowly over time and are usually considered to be benign, and of Grade II, which present nuclear atypia mixed in with the normal brain as cellularity and vascularity is quite low (Batjer and Loftus, 2003). Depending on the cell type they originate from, they can be termed "oligodendrogliomas" (a type of glioma believed to originate from the oligodendrocytes of the brain or from a glial precursor cell), "astrocytomas" (from the star-shaped brain cells in the cerebrum called astrocytes) or mixed type (e.g., oligoastrocytomas, containing cells from different types of glia). On conventional MR images, LGGs present a rather homogeneous structure usually with no contrast enhancement or peritumoral edema (Price, 2010), as illustrated in Figure 9.1.

On the other hand, high-grade gliomas (HGG) consist of Grade III and Grade IV glial tumors. Grade III present mitoses and anaplasia (from Greek: *ana*, "backward" + *plasis*, "formation," meaning with poor cellular differentiation), and their most common subtype is anaplastic astrocytoma (AA) which is illustrated in Figure 9.2. Grade IV gliomas are

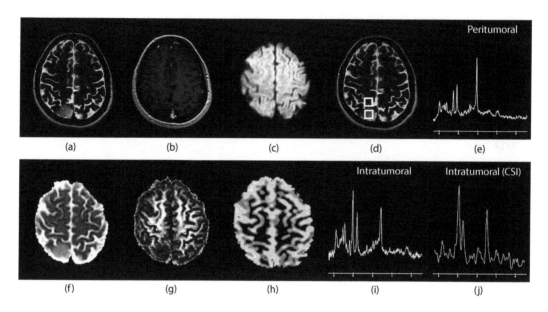

FIGURE 9.1   A case of a low-grade glioma, presenting high signal intensity on a T2-weighted image (a), no contrast enhancement on a T1 3D-SPGR image (b), and an isointense signal on a diffusion-weighted image (c). The lesion shows increased ADC (f), lower FA (g), and no significant perfusion (h) on the corresponding parametric maps. The peritumoral (e) and intratumoral (i and j) spectra are also depicted. The MR spectroscopy voxel's placement is depicted in (d).

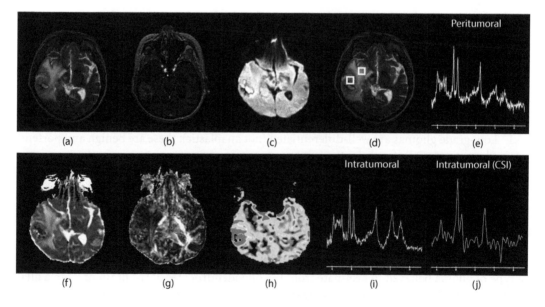

FIGURE 9.2   A case of an anaplastic astrocytoma, presenting high signal intensity on a T2-weighted image with peritumoral edema (a), heterogeneous contrast enhancement on a post-contrast T1 3D-SPGR image (b), and restricted diffusion in the solid portion of the tumor (c). The lesion is hypointense on the ADC map (f), presents isointense FA (g), and increased perfusion on the rCBV map (h). The corresponding peritumoral (e) and intratumoral (i and j) spectra are also depicted. The MR spectroscopy voxel's placement is depicted in (d).

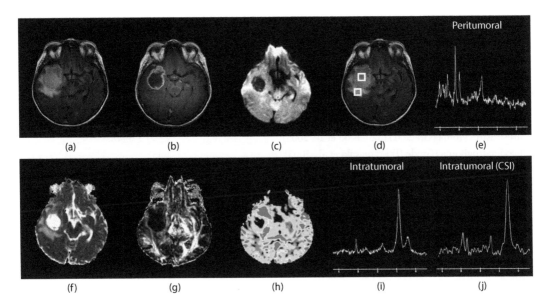

FIGURE 9.3 A case of a glioblastoma multiforme presenting high signal intensity on a T2-weighted image (a) and ring-shaped enhancement on a T1-weighted post-contrast image (b). On the DW-image the lesion presents low signal intensity (c) resulting in higher intratumoral ADC (f), lower intratumoral FA (g), and high peritumoral rCBV (h) reflecting tumor infiltration in the surrounding parenchyma. The corresponding peritumoral (e) and intratumoral (i and j) spectra are also depicted. The MR spectroscopy voxel's placement is depicted in (d).

characterized by increased cellularity and vascular proliferation with extended necrosis on pathologic evaluation, and are usually termed as glioblastoma multiforme (GBM), which is illustrated in Figure 9.3. HGGs present heterogeneous or ring-shaped contrast enhancement patterns, necrotic or cystic areas, hemorrhage and infiltrative edema. A simple grading system for all gliomas can be based on four distinct parameters: nuclear atypia, mitoses, endothelial proliferation, and necrosis. When two or more of the aforementioned features exist, a glioma is categorized as a HGG.

However, the imaging characteristics of the two main glioma categories (LGG vs HGG) are not always grade-specific, and not infrequently, low-grade gliomas present similar morphological features to high-grade gliomas and vice versa (Fan et al., 2006; Liu et al., 2011). It follows that if evaluation is based on conventional MRI alone, these imaging similarities may potentially lead to inaccurate tumor characterization and staging. Hence, a multiparametric approach with advanced MR techniques would provide an insight into the underlying pathophysiology, utilizing parameters and revealing aspects that will allow a more accurate interpretation.

9.2.1 DWI Contribution in Gliomas

The contribution of DWI metrics in the differentiation of lower and higher glioma grades has been studied by several research groups, but with contradicting results and ambiguous conclusions. LGGs usually present higher ADC values compared to HGGs, most probably due to their cellular structure (Yamasaki et al., 2005); however, there can be overlap between the ADC values of both groups. One of the first diagnostic applications of DWI in brain tumors was the differentiation between epidermoid tumors and extra-axial cysts (Tsuruda et al., 1990) since in

TABLE 9.1   Reference ADC Values of Gliomas, Normal Brain and Associated Pathology

Normal Brain Tissue	ADC μm²/ms
Normal white matter (Maier et al., 1998)	0.705 ± 0.014
Deep gray matter (Helenius et al., 2002)	0.75 ± 0.03
HGGs and LGGs	**ADC μm²/ms**
Pilocytic astrocytoma (Grade I) (Yamasaki et al., 2005)	1.659 ± 0.260
Diffuse astrocytoma (Grade II) (Yamasaki et al., 2005)	1.530 ± 0.148
Oligodendroglioma (Grade II) (Yamasaki et al., 2005)	1.455
Anaplastic astrocytoma (Grade III) (Yamasaki et al., 2005)	1.245 ± 0.153
Anaplastic oligodendroglioma (Grade III) (Yamasaki et al., 2005)	1.222 ± 0.093
Glioblastoma (Grade IV) (Yamasaki et al., 2005)	1.079 ± 0.154
Others Brain Tumors	**ADC μm²/ms**
Metastatic tumor (Yamasaki et al., 2005)	1.149 ± 0.192
Typical meningioma (Hakyemez et al., 2006c)	1.17 ± 0.21
Atypical/malignant meningioma (Hakyemez et al., 2006c)	0.75 ± 0.21
Malignant lymphoma (Yamasaki et al., 2005)	0.725 ± 0.192
Other Pathologies	**ADC μm²/ms**
Vasogenic (peritumoral) edema (Maier, Sun, and Mulkern, 2010)	1.30 ± 0.11
Peritumoral edema (HGG) (Morita et al., 2005)	1.825 ± 0.115
Cystic/necrotic tumor area (Chang et al., 2002)	2.70 ± 0.31
Abscess (Chang et al., 2002)	0.65 ± 0.16

cystic lesions, the diffusion coefficient is similar to pure water and can easily be distinguished from that of a tumor. Moreover, necrotic areas may also easily be distinguished, having an overall elevated ADC (see Figure 9.3d, the high intratumoral ADC in GBM). There is a wide range of different diffusion values in the literature, both for tumors and the normal brain, with a study from Yamasaki et al. (2005) being the most comprehensive, including standard deviations (Yamasaki et al., 2005). Based on Yamasaki, Table 9.1 presents the ADC values for the discussed brain tumors, normal brain and associated pathology for reference. Moreover, in order to have a quick reference of the range of ADC values and image signal intensity, Figure 9.4 shows their relative position on an ADC scale.

Coming back to LGGs vs HGGs, Zonari et al. (2007) reported that even though diffusion was higher in LGG, large variations of ADC values existed between the two groups, thus no significant differences were observed. Nevertheless, the exact placement of the region of interest (ROI) area of measurement is always a significant issue that may determine the evaluation of diffusion results. Still, several studies have also concluded that DWI metrics, either from the solid part of the tumor or from the peritumoral edema, are inadequate to provide information about the degree of differentiation of glial tumors (Lam et al., 2002; Pauleit et al., 2004; Rizzo et al., 2009). On the other hand, in the study by Kono et al. (2001), the difference of ADC values between glioblastomas and grade II astrocytomas reached statistical significance, however the authors reported that peritumoral neoplastic cell infiltration cannot be revealed using individual ADC values or even by evaluating ADC maps. In the same study, an inverse relationship was observed between diffusion and tumor cellularity, where lower ADC values

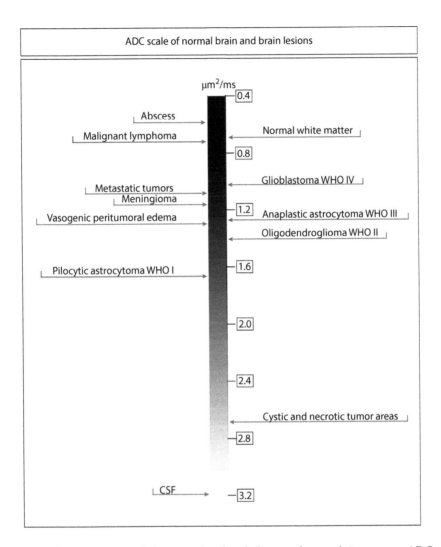

FIGURE 9.4   Relative positions of gliomas, related pathology, and normal tissue on an ADC scale.

suggested malignant gliomas, whereas higher ADC values suggested low-grade astrocytomas (see Figure 9.1f, vs. Figure 9.2f). Nevertheless, the authors concluded that even though ADC values cannot be reliably used in individual cases to differentiate tumor type, a combination of routine image interpretation and ADC results in a higher diagnostic value. Keep in mind that very high ADC values can be depicted in HGGs but this would be the result of extended necrosis (Figure 9.3f). On the other hand, it has been shown that DWI metrics may be useful in the differentiation of non-enhancing gliomas, as ADC values in anaplastic astrocytomas can be significantly lower in the solid portions of the tumors compared to LGG (Fan et al., 2006). Last, a very recent study by Ryu et al. also confirmed the differentiation between LGGs and HGGs using measures of ADC texture, confirming the hypothesis that diffusion value in tumors is inversely correlated with cell density and nucleus/cytoplasm ratio (Ryu et al., 2014).

In conclusion, DWI does certainly aid in the diagnostic decision of gliomas, although the considerable overlap of ADC values does not permit the diagnosis of a particular brain tumor. Nonetheless, DWI may also aid in the prediction of treatment outcome and ADC texture characteristics may hold prognostic information (Brynolfsson et al., 2014). Such methods are usually referred to as "functional diffusion imaging," since ADC values are measured over time and compared on a coregistered voxel-by-voxel basis.

9.2.2 DTI Contribution in Gliomas

The ability of DTI parameters to discriminate between LGG and HGG by assessment of tumor cellularity is also a matter of great dispute in the current literature. Some studies have reported that mean diffusivity (MD) within grade I gliomas was significantly higher compared to grade III and grade IV gliomas, respectively, but no differences were observed in MD values between grade II and grade III gliomas (Inoue et al., 2005). In contrast, it has also been reported that MD cannot be used as a predictor of lower and higher glioma grades either in the intratumoral area (Stadlbauer et al., 2006; Tropine et al., 2004; Wang et al., 2011) or in the peritumoral area (Toh et al., 2008a; Wang et al., 2011). However, Lee et al. (2008) showed that MD is significantly lower in the non-enhancing regions of HGG compared to LGG although HGG may present relatively benign imaging findings, such as absence of contrast enhancement. Similar tendencies were observed in other studies as well, but to date, no statistical significance has been reached, probably because of MD's dependence on tumor cellularity, intra- or extracellular edema, and tumor necrosis, at least at the typical b value of 1000 s/mm^2 (Goebell et al., 2006; Liu et al., 2011).

Although the DTI related metrics have not yet been shown to correlate with tumor cellularity, fractional anisotropy (FA) has also been investigated regarding glioma grading and tumor infiltration. Studies of the relationship between DTI and histological malignancy of gliomas showed that FA can distinguish HGG from LGG, thus it can contribute to the surgical strategy decision or the selected site of stereotactic biopsy (Inoue et al., 2005; Lee et al., 2008; Stadlbauer et al., 2006; Tropine et al., 2004).

Inoue et al. reported that the FA value could be used to distinguish HGG from LGG since FA is significantly higher in HGG than LGG and that high anisotropy implies that the tissue is symmetrically organized. However, these results are somewhat contradictory to the usual understanding of the microstructure of high-grade gliomas. HGGs typically depict a regressive organization rather than an increase in symmetric histological organization. A positive correlation of FA with cell density of gliomas was established and a FA cut-off value of 0.188 was determined between the two groups (Inoue et al., 2005).

Nonetheless, the correlation between FA and glioma cellularity has also been disputed. Although Stadlbauer et al. (2006) reported that FA is a better indicator than MD for the assessment and delineation of different degrees of pathologic changes in gliomas, they concluded in a negative correlation between FA and glioma cellularity. In agreement with their non-infiltrating nature, the presence of well-preserved fibers in the periphery of LGG, would suggest higher FA values in contrast to HGG, where peritumoral tracts are disarranged or disrupted (Chen et al., 2010; Goebell et al., 2006; Svolos et al., 2014). The evaluation of FA in the peritumoral area has been reported to provide useful information and differentiation from LGG even in gliomas with no contrast enhancement (Liu et al., 2011). Although the determination of a specific cut-off value is difficult, Liu et al. observed that the mean and maximal FA values were significantly lower in LGG, and proposed a cut-off value of 0.129 between the two groups. They also proposed that the combination of these two parameters improves diagnostic accuracy; therefore, it may be useful in the preoperative grading of non-enhancing gliomas. Moving forward, Ferda et al. (2010) reported that although the evaluation of the FA maps might not be sufficient for glioma grading, the combination of FA metrics with contrast enhancement pattern improves the possibility of distinguishing LGG from HGG.

Very recently, Server et al. (2014) conducted an analysis of diffusion tensor imaging metrics for gliomas grading at 3T and evaluated the diagnostic accuracy of axial diffusivity (AD), radial diffusivity (RD), apparent diffusion coefficient (ADC) and FA values. They concluded that these are useful DTI parameters for differentiation between low- and high-grade gliomas and reported a diagnostic accuracy of over 90%.

Nevertheless, controversies regarding the contribution of FA still exist in the literature, as a number of studies have reported that the utility of DTI metrics in grading gliomas is still limited, especially taking into account the potential artifacts and pitfalls of the method (Chen et al., 2010; Lee et al., 2008; Wang et al., 2011).

9.2.3 Perfusion Contribution in Gliomas

The evaluation of perfusion metrics using dynamic contrast enhancement (DSC) has been shown to provide a robust differentiation between LGG and HGG contrary to diffusion techniques. Important histopathological factors like vascular morphology and angiogenesis, which determine the degree of malignancy and grade of glial tumors, can be reflected in the estimated rCBV values (Cha, 2004; Hakyemez et al., 2005). LGGs present no or minimally increased rCBV in the intratumoral area compared to the contralateral normal side, due to their low vascularity, and significantly lower mean rCBV when compared to HGGs. More specifically, anaplastic astrocytomas present with higher rCBV values than LGGs but lower than the most hypervascular GBMs.

DSC-based studies have been generally successful in the differentiation between LGGs and HGGs based on the characteristic difference of the underlying vascularity between the two tumor types (Di Costanzo et al., 2008; Rizzo et al., 2009; Server et al., 2009; Zonari et al., 2007).

Although, a range of mean rCBV values have been assigned to each glioma group, there is an agreement regarding the correlation between microvascular density and glioma grade. In other words, all studies agree that moving from lower to higher glioma grades, there is a progressive increase in rCBV values (see Figure 9.1h vs. Figure 9.2h vs. Figure 9.3h).

One exception that has to be taken into account when glioma-grade comparisons are conducted is the case of low-grade oligodendrogliomas and pilocytic astrocytomas. Although considered benign, they have been shown to exhibit increased angiogenesis and elevated rCBV values that may exceed the calculated threshold of HGGs (Cha et al., 2005; Law et al., 2008; Lev et al., 2004). Moreover, it should be noted that since gliomas consist of a relative heterogeneous group of tumors, there can be an overlap in rCBV values among different grades.

Last, the comparison between non-enhancing HGG and LGG in terms of perfusion measurements has yielded contradicting results, as Liu et al. (2011) did not establish differences in the rCBV ratios between the two groups, whereas Fan et al. (2006) and Maia et al. (2005) reported that rCBV values from the non-enhancing regions of HGG were significantly higher than LGG, but no threshold was proposed between the two.

Despite the relative variations in the rCBV values observed, the overall conclusion is that rCBV presents, in general, a strong and positive correlation to glioma grade, thus it may constitute an important marker of tumor angiogenesis and malignancy.

Perfusion can be also studied with DCE MRI as analytically explained in Chapter 3. Although the studies of brain tumors with DCE are fewer in number, there is recent evidence with promising results in the characterization of tumors and good correlation between DCE metrics and tumor grade (Bauer et al., 2015; Jung et al., 2015; Li et al., 2015). More specifically, fractional vascular volume and brain tumor grade were adequately correlated, while differentiation of LGGs from HGGs was accomplished using the K^{trans} and v_e

parameters (Choi et al., 2013; Li et al., 2015). It has to be mentioned here that although some reports suggest that DSC measurements like rCBV may be slightly more predictive than DCE, both measures are strongly correlated in HGG, suggesting that some combination of both may prove ideal (Law et al., 2006; Provenzale et al., 2006).

9.2.4 MRS Contribution in Gliomas

The clinical use of proton MRS for brain tumor evaluation depends on the existence of a relationship between the evaluated metabolites and specific clinical features including histological category, grade, location, malignancy, etc. Nevertheless, spectra may be highly variable, even in tumors of the same type and histological grade (Howe et al., 2003), and this cause for variability is still an area of active research. A review of the literature demonstrates that MRS is increasingly used in clinical studies to non-invasively identify regions with metabolic specific characteristics that reflect glioma type and grade. More specifically for glial tumors, a common observation is elevation in choline (Cho), lactate (Lac) and lipids (Lip), with a significant decrease in N-acetyl aspartate (NAA). As a general rule, the higher the glioma grades the higher Cho, Lac, and Lip and the lower NAA and Cr (see Figure 9.1i vs. Figure 9.2i vs. Figure 9.3i). Nevertheless, there can be a significant overlap among different glioma types (Howe et al., 2003; Soares, 2009) and therefore, MRS is currently used primarily to differentiate glial tumor grade rather than to confirm a histologic diagnosis.

Differentiation between LGGs and HGGs and grading is an important clinical issue, especially taking into account the dispute on the optimum treatment strategy for patients with low-grade tumors. Although it has been shown that differences in Cho levels can provide differentiation with high sensitivity and specificity (Law et al., 2003), it remains an open question whether 1H-MRS is able to define WHO grade of gliomas. However, a recent study by Porto et al. (2011) revealed a more prominent loss of NAA and increase of Cho in WHO III than in WHO II astrocytomas. NAA/Cho ratio was identified among the other ratios as the most accurate to discriminate between tumor grades. This is in agreement with the general consensus that NAA/Cho ratios decrease with higher histological grade of gliomas. Similarly, Chiang et al. (2004) revealed significantly higher Cho/NAA and Cho/Cr ratios in high-grade astrocytomas than in the low-grade astrocytomas. Law et al. (2003) demonstrated a threshold value of 1.56 for tCho/tCr ratio with 75.8% sensitivity and 47.5% specificity for the determination of HGG versus LGG. Moreover, a threshold value of 1.6 for tCho/NAA provided 74.2% sensitivity and 62.5% specificity in predicting the presence of a HGG (Law et al., 2003). It follows that there is a consistent correlation between Cho increase - NAA decrease and tumor grade. A controversy is that lower average Cho has been observed for glioblastomas when compared to anaplastic astrocytomas (WHO III), which can be explained as a dilution of Cho by the presence of extensive necrosis (Howe et al., 2003).

In short TE spectra (e.g., 35 ms), large signals from mobile lipids can be considered an indicator of grade and prognosis in patients with glial tumors (Figure 9.3i) (Kimura et al., 2001). In fact, a research group demonstrated that instead of Cho, the amount of lipids can be used as the second-best discriminator between LGG and HGG, with glioblastomas exhibiting the highest amount of lipids since necrosis is one of their microscopic hallmarks (Möller-Hartmann et al., 2002). Although it has been previously proposed that lactate also increases with grade, it cannot be safely used as a discriminator metric. The most probable explanation for that is the difficulty of accurately quantifying lactate in the presence of high lipid signals.

Furthermore, studies using short TE have shown that mI levels may aid tumor classification and grading (Castillo et al., 2000; Hattingen et al., 2008; Howe et al., 2003; Law et al., 2002).

More specifically, LGGs express higher levels of mI compared to HGGs. Regarding glutamine and glutamate (Glx), it has been shown that they can be used for the distinction between oligodendrogliomas and astrocytomas. Rijpkema et al. (2003) found significantly increased Glx levels for oligodendrogliomas when compared to that of astrocytomas in short TE spectra.

Despite the aforementioned results, the clinical utility of proton MRS for glioma differentiation and grading is still questioned, mainly due to the high variability as well as the significant overlap of spectra. Consequently, although MRS is very sensitive to abnormal metabolic changes, the specificity remains relatively low. Nevertheless, one of the most interesting results in the study by Server et al. (2010) was the elevation of Cho/Cr and Cho/NAA metabolite ratios in the peritumoral region, which was correlated to glioma grading. Therefore, the peritumoral region can be evaluated with MRS under the strong hypothesis that higher grade gliomas are infiltrating intracerebral tumors, hence regions of altered signal outside the enhancing margins of LGGs represent a variable combination of vasogenic edema and infiltrating tumor cells. A great deal of evidence suggests that the predominance of tumor cells produces spectra with the typical tumor pattern (i.e., with high Cho and low NAA peaks, and abnormal Cho/NAA ratios) (Di Costanzo et al., 2008). On this basis, significantly lower NAA/Cr of both TE and higher long TE Cho/Cr and Cho/NAA ratios in the peritumoral area of HGGs when compared to LGGs are found (Kousi et al., 2012). As mentioned earlier though, if edema is prominent, the increase of interstitial water "dilutes" the signal of metabolites and therefore there can be cases where HGGs will not reveal evidence of infiltration.

Moving forward, Scarabino et al. (2009) analyzed the metabolite spatial distribution in the peritumoral area of gliomas and they successfully discriminated LGGs from HGGs. Similarly, Server et al. (2010) observed elevation in peritumoral Cho/Cr and Cho/NAA metabolic ratios related to grading. The peritumoral Cho/NAA and Cho/Cr ratios were significantly higher in high-grade than in low-grade gliomas.

In conclusion, as illustrated in one of our studies that included 71 patients with LGGs and HGGs, the peritumoral area of gliomas proved to be more valuable in predicting glioma grade as more metabolic ratios (four ratios) significantly differentiated the two tumor groups over the intratumoral area (three ratios) (Kousi et al., 2012). This is illustrated in Figure 9.5. Therefore, it is suggested that the peritumoral area should always be included in the MRS evaluation.

Lastly, low-grade neoplasms have also been evaluated to be prone to malignant degeneration (Wu et al., 2002). This can be detected using MRSI as increased choline and discrepant spectroscopic results, and may be used to reorient the treatment strategy based on an earlier disease progression.

9.3 Cerebral Metastases

Focus Point 'Advanced techniques in Cerebral Metastases'

DWI
- Direct differentiation of secondary tumors from low-grade gliomas using ADC.
- Conflicting results in the literature regarding differentiation ability between metastases and high-grade gliomas.

DTI
- Cerebral metastases present decreased FA values in the intratumoral and peritumoral region compared to high-grade gliomas.

PWI
- rCBV does not provide robust differentiation between metastases and HGG
- Peritumoral rCBV higher in HGG compared to metastases.

MRS
- Intratumoral spectra of metastases and HGG almost identical.
- Differentiation of metastases vs. HGG, using peritumoral Cho/Cr, Cho/NAA, and minimum NAA/Cr.

fMRI
- Localization of brain function for pre-surgical planning.

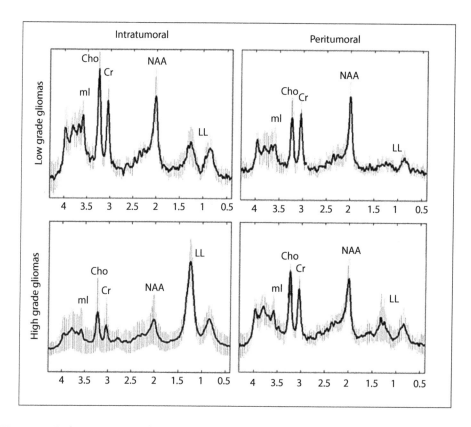

FIGURE 9.5 Indicative in vivo short TE mean spectra with their corresponding standard deviations (vertical lines) from the intratumoral and peritumoral areas of 19 low-grade and 52 high-grade gliomas.

Intracranial metastases are the most common brain tumors in adults and account for approximately 25%–50% of intracranial tumors in hospitalized patients with their incidence rapidly increasing. The primary cancers that are most likely to metastasize to the brain are lung, breast, colon, malignant melanoma and gastro-intestinal cancers (Sawaya, 2001; Patchell, 1991). Of course, there is the unknown primary tumor as well. The majority of metastatic tumors will be parenchymal; fewer will be located to the subarachnoid space, and rarely to other structures such as choroid plexus or the pituitary gland.

Metastases differentiation from other malignant tumors on conventional MRI is usually a straightforward process, due to the clinical history of the patient or the existence of multiple well-circumscribed lesions seen in 60%–75% of the cases. However, the existence of a solitary enhancing lesion may complicate differential diagnosis because it may present similar imaging characteristics and contrast enhancement patterns like those of HGGs (see Figure 9.3 vs. Figure 9.6). Hence, advanced neuroimaging techniques have the potential of increasing the

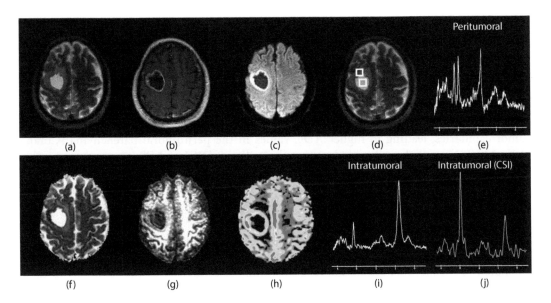

FIGURE 9.6 A case of an intracranial metastasis presenting high signal intensity on a T2-weighted image (a), ring-shaped enhancement on the T1-weighted post-contrast image (b), restricted diffusion in the periphery of the tumor (c), increased intratumoral ADC (f), decreased FA (g), and elevated perfusion in the peripheral solid part of the lesion (h). The corresponding peritumoral (e) and intratumoral (i and j) spectra are also depicted. The MR spectroscopy voxel's placement is depicted in (d).

sensitivity and specificity of metastasis detection, thus substantially improving surgical planning and therapeutic approach (Lee, 2012).

9.3.1 DWI/DTI Contribution in Metastases

Solitary brain metastases and primary high-grade gliomas represent one of the most common differential diagnostic problems in the clinical routine and thus it has been extensively investigated in the literature. The differentiation of these two very common brain tumors depends on the following hypothesis.

Metastatic brain tumors arise mostly within the brain parenchyma and usually grow by expansion, displacing the surrounding brain tissue, with no histologic evidence of tumor cellularity outside the contrast-enhanced margin of the tumor (pure vasogenic edema) (Bertossi et al., 1997). On the other hand, HGGs, as already mentioned, are characterized by the ability to recruit and synthesize vascular networks for further growth and proliferation and therefore tumor cells are expected to be present in their periphery along with increased edema concentration (infiltrating edema).

Both metastases and high-grade gliomas have been reported with increased ADC and lower FA values intratumoraly due to their heterogeneous cellular structure. In the peritumoral edematous region, similar ADC and FA values have been reported, despite the fact of their different nature (vasogenic versus infiltrating). Hence, because of the aforementioned similarities, a large number of studies in the literature have concluded that the contribution of DWI and DTI metrics, either in the tumor or the peritumoral area, is still limited (Lee, 2012; Pavlisa et al., 2009; Tsougos et al., 2012; Wang et al., 2009; Yamasaki et al., 2005).

Contrary to the aforementioned observations, a number of studies report that the diffusion profiles of HGGs and metastases differ and that DWI/DTI measurements may be indicative for tumor discrimination, introducing the "tumor infiltration index" (Chiang et al., 2004; Lu et al., 2004). Based on these studies, HGGs present elevated FA in the intratumoral and peritumoral region compared to metastases, whereas the latter present increased ADC values in their periphery. Higher cellularity in the solid part and presence of tumor cells in their periphery due to the infiltrating nature of HGGs, might explain the restricted diffusion compared to metastases (Altman et al., 2007; Gillard et al., 2005). On the other hand, metastases have been associated with increased edema concentration, as a result of their leaky tumor capillaries, leading to higher ADC values in the peritumoral parenchyma (Chiang et al., 2004). Nevertheless, it is evident that the utility of DWI and DTI in preoperative differentiation of solitary metastatic tumors from high-grade gliomas remains controversial. Our studies have shown that based solely on the mean values and the corresponding SDs of FA and ADC measurements (Figure 9.7), the mean value of FA in glioblastomas is higher than in metastases, indicating roughly that FA might be a more appropriate index to quantify the diffusion properties in the intratumoral region of these two groups (Tsougos et al., 2012). The higher FA of glioblastomas in this patient cohort (35 GBM and 14 Metastases) may be attributed to the fact that glioblastomas present higher cellularity in the solid part of their lesion than do brain metastases (Altman et al., 2007). Whether FA is positively or negatively

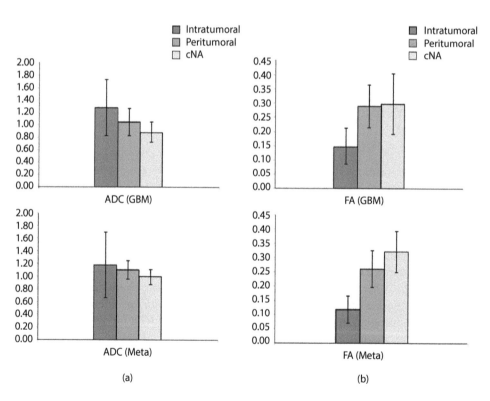

(a) (b)

FIGURE 9.7 Histograms illustrating a comparison of ADC (a) and FA (b) measurements in terms of the mean values with their corresponding S.D. in the intratumoral region (upper row), peritumoral region (lower row) and the contralateral normal area (cNA) for glioblastomas (GBMs) and metastases (Meta). (Copyright © I. Tsougos et al. 2012, International Cancer Imaging Society.)

correlated with tumor cell density and vascularity, and whether it can be used to assess tumor grading, is still under wide dispute.

9.3.2 Perfusion Contribution in Metastases

Based on the hypothesis of vasogenic versus infiltrating edema around metastases and HGGs respectively, increased rCBV ratios have been measured in the surrounding edema of HGGs compared to solitary metastases, enabling the differentiation of these two tumor groups, although wide variability with considerable overlap has also been reported (Hakyemez et al., 2010; Tsougos et al., 2012). Regarding their intratumoral region, no differences have been observed in the measured rCBV ratios (both demonstrate increased rCBVs, Figure 9.3h vs. Figure 9.6h), most probably due to high vascularity and abnormal capillary permeability of both tumor types (Chiang et al., 2004; Hakyemez et al., 2010; Tsougos et al., 2012). More recently, Jung et al. (2015) suggested that the shape of the enhancement curve can also distinguish between glioblastoma and hypovascular metastasis (Jung et al., 2015). Nevertheless, DCE permeability analysis showed no statistical significance in the distinction of the two tumor groups.

9.3.3 MRS Contribution in Metastases

MRS in cerebral metastasis shows low or absent NAA, which is consistent with the lack of neuroglial elements, elevated choline/creatine ratio and elevated lactate and lipids as in HGGs since anaerobic glycolysis and tissue necrosis are common features of both tumors (see Figure 9.3 vs. Figure 9.6). Hence, the MRS characteristics of the intratumoral areas of these two lesions are identical.

Metastatic lesions were initially thought to be distinguishable from high-grade gliomas by identifying high degree of lipid and lactate signals and increased lipid/creatine ratio (Bulakbasi et al., 2003; Ishimaru et al., 2001). Especially regarding mobile lipids, Opstad et al. (2004) detected two different types of lipid and macromolecule signals and tried to investigate the usefulness of their relative difference between HGGs and metastases. They concluded that highly mobile (with very sharp clearly defined peaks, L1) and less mobile lipids (with broad, overlapping peaks, L2) could differentiate between the two lesion types. The ratio L1/L2 was found to be 2.6 in HGG and 3.8 in metastasis ($p < 0.0001$) allowing a discrimination between the two, with sensitivity and specificity of 80%. However, since both lesions contain areas of extended necrosis, the specificity of the above findings is decreased.

Therefore, again based on the vasogenic vs infiltrating edema hypothesis, interest was focused in the peritumoral area. The hypothesis here is that glioblastomas are infiltrative lesions, thus pathological tissues may be present in the surrounding white matter, while not in metastatic lesions. Hence, an increased choline/creatine ratio in the periphery of the lesion can sometimes distinguish HGGs from metastasis. Based on these peritumoral characteristics, MRS studies have shown significant distinction between the two lesions (Chawla et al., 2010; Chiang et al., 2004; Tsougos et al., 2012). Figures 9.3e and 9.6e illustrate how the difference in the peritumoral spectra can differentiate HGGs from metastases. The peritumoral area of the GBM presents increased Cho/Cr ratio and decreased NAA, while the peritumoral spectrum of metastasis only shows mildly decreased NAA.

9.4 Meningiomas

Focus Point

DWI/DTI

- Inverse relationship between water diffusion and malignancy.
- Lower ADC and higher FA values for atypical/malignant meningiomas compared to benign.
- FA is significantly different between subtypes of benign meningiomas, whereas ADC cannot contribute.

PWI

- The simple time–intensity curve (TIC) is one of the most useful perfusion parameters in the case of meningiomas.
- Characteristic early enhancement during the arterial phase remains after the venous phase.

MRS

- Ala and Glx peaks can be characteristic for meningiomas (but appear only in 30–40% of cases).
- Differentiation of atypical meningioma vs. HGG, using peritumoral Cho/Cr, Cho/NAA and minimum NAA/Cr.

fMRI

- Localization of brain function for pre-surgical planning.

A meningioma is a tumor that arises from the meninges, that is the membranous layers surrounding the central nervous system. Although not technically a brain tumor (it is situated on the brain), it is included in this category because it may compress or squeeze the adjacent brain. Therefore, meningiomas are the most common extra-axial cerebral tumors, and due to their characteristic location a relatively straightforward diagnosis is achieved. However, although the majority of meningiomas are benign, they can have malignant transformations. According to the WHO classification system, there are three types of meningiomas. The majority of meningiomas (up to 90%) are benign or Grade I and usually full recovery is achieved with surgical resection. Grade II (atypical) and Grade III (malignant) meningiomas are less common but more aggressive than Grade I, thus they are more likely to recur even after complete resection. The differences between benign and atypical/malignant meningiomas relate to the number of mitoses, cellularity, and nucleus-to-cytoplasm ratio as well as their histologic patterns (Perry et al., 2007). Conventional MR imaging provides useful information regarding their localization and morphology, however there can be cases where meningiomas may have atypical imaging findings, like heterogeneous contrast enhancement and necrotic areas mimicking high-grade tumors. Hence, a correct diagnosis and accurate histologic grading is of great importance for beneficial treatment planning. An atypical/malignant meningioma with a large heterogeneous enhancement and intense mass effect is illustrated in Figure 9.8.

9.4.1 DWI/DTI Contribution in Meningiomas

The usefulness of DWI/DTI techniques in meningioma grading or in the differentiation of benign and malignant subtypes has been previously systematically investigated. Studies

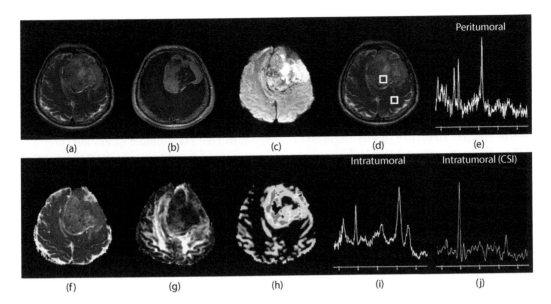

FIGURE 9.8 A case of an atypical meningioma. Axial T2-weighted (a) and postcontrast T1-weighted (b) images demonstrate a large heterogeneous enhanced left frontal mass. The lesion presents areas of restricted diffusion (c), isointensity on the ADC map (f), hypointensity on the FA map (g), and elevated rCBV (h). The corresponding peritumoral (e) and intratumoral (i and j) spectra are also depicted. The MR spectroscopy voxel's placement is depicted in (d).

have shown (Filippi et al., 2001; Hakyemez et al; 2006c; Nagar et al., 2008; Toh et al., 2008a) that there is an inverse relationship between water diffusion and malignancy, with lower ADC and higher FA values for malignant meningiomas compared to benign. It seems that the restricted water diffusion reflects the increased mitotic activity and necrosis as well as the high nucleus-to-cytoplasm ratio present in high-grade meningiomas (Buetow et al., 1991). On the other hand, benign meningiomas are characterized by a more isotropic diffusion since they consist of oval and spindle-shaped neoplastic cells forming fascicles, cords, or nodules (Ellison et al., 2004). For the differentiation between subtypes, FA was significantly different between subtypes of benign meningiomas (Tropine et al., 2007), whereas ADC did not contribute either in benign or malignant subtype discrimination (Ginat et al., 2010; Tropine et al., 2007). However, as in the case of gliomas, controversies still exist in the literature regarding meningiomas, as a number of previous studies conclude that diffusion quantification, derived either from the tumor (Pavlisa et al., 2008; Santelli et al., 2010), or from the peritumoral edema (Hakyemez et al., 2006c; Toh et al., 2008a), cannot provide significant information for meningioma grading.

Tropine et al. (2007) found significant differences in MD and FA between meningiomas and LGGs. LGGs had increased MD and decreased FA values in the intratumoral region compared to meningiomas, probably due to their lower tumor cellularity. No differences were observed in the related parameters of the peritumoral edema, which makes sense due to the non-infiltrating nature of both lesions.

The contribution of DWI/DTI metrics is more important in the differentiation between atypical meningiomas and HGGs. Meningiomas that exhibit atypical MRI features such as cystic and necrotic areas, ring-like enhancement and parenchymal invasion, may resemble gliomas or metastatic brain tumors leading to false radiological reports and misinterpreted treatment decisions. The incidence of such cases can rise up to 15%, constituting a serious differential diagnostic problem.

Again, the results in the literature are conflicting. Several studies have suggested that diffusion metrics can be helpful in correctly characterizing these lesions to optimize treatment planning. The main finding is lower ADC and higher FA values in the solid parts of meningiomas as compared to HGG (De Belder et al., 2012; Hakyemez et al., 2006; Tropine et al., 2004). This behavior suggests a higher level of fibrous organization and a more incoherent cellular structure in meningiomas compared to HGG (De Belder et al., 2012).

As already mentioned atypical meningiomas may resemble solitary metastatic tumors as well. The diffusion profiles in meningiomas and metastases present great similarities that have also been reported by several groups (Lu et al., 2004; Van Westen et al., 2006; Wang et al., 2011). Based on these studies, diffusion and anisotropy changes, either from the solid region or the periphery of the tumor, are inadequate to distinguish meningiomas from metastases. These findings may be explained by the fact that atypical/malignant meningiomas often present a heterogeneous cellular structure, with necrotic and cystic portions, thus inducing unhindered water diffusion comparable to that of metastatic tumors. Furthermore, as non-infiltrating lesions, their surrounding edema is purely vasogenic, and cannot provide distinct information in terms of DWI/DTI measurements.

9.4.2 Perfusion Contribution in Meningiomas

Meningiomas are highly vascular lesions that primarily derive blood from meningeal arteries, and as such, they demonstrate very high perfusion, regardless of their grade. Another characteristic is their increased contrast leakage and permeability, which is observed on perfusion images due to the complete lack of the capillaries in the blood-brain barrier (BBB). The simple time–intensity curve (TIC) is one of the most useful perfusion parameters in the case of meningiomas; it reveals a characteristic large and rapid drop in signal intensity during the first pass of contrast relative to the normal brain, which will then slowly return to a level lower than the parenchyma, as illustrated in Figure 3.1 (please see Chapter 3). This characteristic (early enhancement during the arterial phase and well after the venous phase) is sometimes jokingly named: "the mother-in-law sign," after the uncanny ability of mothers-in-law to arrive early and stay late. Despite the probably "unkind" description, this characteristic is believed to reflect the increased size, number, or tortuosity of vessels in meningiomas compared to the normal brain as well as the lack of BBB (Collins and Christoforidis, 2016). DCE metrics may also contribute in separating typical from atypical meningiomas, using the volume transfer constant, a measure of capillary permeability (Yang et al., 2003).

Perfusion MRI may also be useful in meningioma grading. For example, Zhang et al. (2008) reported that malignant meningiomas had higher rCBV ratios in their periphery compared to benign. However, in intratumoral rCBV measurements, there was no statistical difference between the two types, most probably due to their inherent hypervascularity (Zhang et al., 2008). Regardless of the high rCBV values of meningiomas, a distinct differentiation from HGGs based solely on rCBV measurements has not been reported (Sentürk et al., 2009). This might be attributed to the highly leaky and permeable capillaries of meningiomas, which might lead to over- or underestimation of rCBV (Cha, 2004).

Apart from HGGs, atypical meningiomas may be misdiagnosed as solitary metastatic tumors. In such cases intratumoral rCBV measurements may provide significant differentiation among the two tumor groups since metastases have been found to have significantly lower rCBV values, despite their increased vasculature (Hakyemez et al., 2006a,b; Kremer et al., 2004). There also appears to be adequate grading by the use of K^{trans} since higher K^{trans} have been reported in atypical meningiomas, indicative of micronecrosis (Berger, 2003). Last, it is not a surprise that angiomatous type meningiomas present increased CBV compared to fibrous meningiomas.

9.4.3 MRS Contribution in Meningiomas

MRS studies of meningiomas include decreased Creatine and decreased or absent N-Acetyl Aspartate (NAA) reflecting the non-neural origin of the lesion, while Alanine (Ala), Choline (Cho), and Glutamate-Glutamine complex (Glx) are usually elevated (Sibtain et al., 2007). Unfortunately, the characteristic high Ala peaks are not found in all meningiomas (only 30%–40% of cases) and seem to correlate inversely with necrosis within these tumors (Castillo and Kwock, 1999). Moreover, Ala and Glx complex evaluation has certain limitations and are not always easy to evaluate during clinical practice, especially in low field MRI scanners (<3T) (Howe et al., 2003). Lac and Lip can be associated with non-benign tumors, and therefore atypical meningiomas, since they indicate intratumoral hypoxia and micronecrosis, respectively. Lipids have been established as indicators of HGGs and metastatic brain tumors, but the question is whether they can be used to indicate malignancy in meningiomas as well (Buhl et al., 2007). Qi et al. (2008) concluded that lipids may be used as a useful marker in the evaluation of intracranial meningiomas grading. However, those findings are still controversial in the literature and a limited number of studies have elaborately investigated lipids in meningioma diagnosis. The main problem is that lipids and lactate do not always represent an index of micronecrosis and therefore they cannot be taken as proof of meningioma malignancy. Especially regarding the usual positioning of meningiomas (i.e., closely proximal to the skull and mastoid), the spectrum might be unavoidably contaminated by lipids due to the "lipid contamination" phenomenon (see Chapter 6). Furthermore, as described by Fountas et al. (2000) the differentiation of metastases and meningiomas is very difficult because of spectral similarities referred to the absence of NAA and low concentrations of Cr. Nevertheless, again based on the peritumoral area hypothesis, interest can be focused in the peritumoral area to distinguish HGGs from atypical meningioma. Hence, an increased choline/creatine ratio in the periphery of the lesion can suggest HGGs rather than meningioma (see Figure 9.8e vs. Figures 9.2e and 9.3e).

Last, in one of our studies, we evaluated the presence of a high peak at 3.8 pmm on short TE spectra, which may be used as a distinct metabolic feature for the differentiation of meningiomas among other cerebral lesions (Kousi et al., 2012). Thus, as Ala, which is considered another characteristic metabolite for meningiomas, is not consistently detected, this peak at 3.8 ppm can be also considered relatively specific for the recognition of meningiomas. It is hypothesized that this peak might correspond to Glx-a peak or Glx together with glutathione (Yue et al., 2008), as Majós et al. (2004) also demonstrated elevation of Glx-a in meningiomas compared to other brain masses.

9.5 Primary Cerebral Lymphoma

DWI/DTI
- PCLs are highly cellular tumors, hence with very high DWI and low ADC.
- PCLs have been reported with significantly lower ADC and FA values than HGGs.

PWI
- Neovascularization is absent in PCLs—Lower rCBV ratio compared to HGGs.
- Time–intensity curve (TIC) presents very high first-pass leak with minimal or no return to baseline.

MRS
- PCNSL spectrum is characterized by increased Cho, Lac, and lipids, similar to HGGs
- Differentiation of PCNSL vs. metastases, using peritumoral Cho/Cr, Cho/NAA, and minimum NAA/Cr.

fMRI
- Localization of brain function for pre-surgical planning.

Primary cerebral lymphomas (PCLs) are primary intracranial tumors with increased incidence in immunocompetent or immunocompromised patients (Surawicz, 1999). MR imaging of lymphomas classically shows round or oval lesions surrounded by peritumoral edema. The peritumoral area of PCLs may not be clear since they are usually infiltrative and not encapsulated (Go et al., 2006). Nevertheless, PCLs present a remarkable contrast enhancement on conventional MR images due to the complete absence of BBB similar to meningiomas (Figure 9.9b). Although PCLs present a very unique histopathologic characteristic, which is the multiple thick layers of neoplastic cells forming around blood vessels, neoangiogenesis is not a prominent feature (Bataille et al., 2000). However, based on conventional MRI findings alone, distinction of high-grade glioma and solitary metastases from CNS lymphoma is often difficult, because of their diffuse infiltrative growth (Hakyemez et al., 2006b).

9.5.1 DWI/DTI Contribution in PCLs

The differentiation of PCL from other CNS lesions is probably the case where DWI has its greatest value. As highly cellular tumor, PCL has a relatively decreased amount of extracellular space, causing a restriction to free water diffusibility, therefore it has been found to present hyper-intense signal intensity on DWI with significantly lower ADC values than HGGs and metastases (Figure 9.9c and f) (Guo et al., 2002; Toh et al., 2008b). In addition, lower FA values have been also observed in PCLs compared to HGGs (Toh et al., 2008b; Wang et al., 2011), although this can be considered a conflicting finding taking into account PCLs' high cellularity.

9.5.2 Perfusion Contribution in PCLs

As already mentioned, in contrast to HGGs, neovascularization is absent in PCNSLs. Therefore, densely contrast-enhanced PCLs could be distinguished from HGG and metastases on the

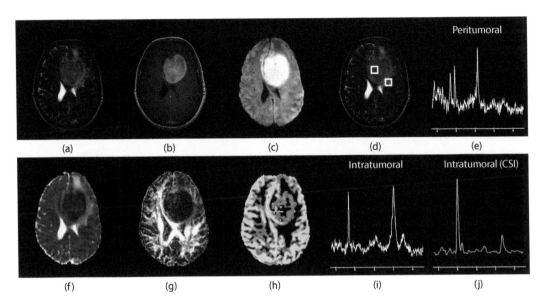

FIGURE 9.9 A case of a primary cerebral lymphoma presenting high signal intensity with peritumoral edema on a T2-weighted image (a) and intense contrast-enhancement on a T1-weighted post-contrast image (b). The lesion is hyperintense on the DW image (c), which results in a hypointense appearance on the ADC (f) and FA (g) maps. The rCBV map shows moderate perfusion within the lesion (h). The corresponding peritumoral (e) and intratumoral (i) and (j) spectra are also depicted. The MR spectroscopy voxel's placement is depicted in (d).

basis of lower rCBV ratio values (Bendini et al., 2011; Hartmann et al., 2003; Lee et al., 2010). Again, although the strong contrast enhancement without CBV increment can be considered a conflicting finding, it is attributed to the BBB destruction without neovascularization, contrary to the increased vascularity in HGG and metastatic tumors (Cha et al., 2002). Furthermore, other researchers also found a time-signal intensity curve characterized by an increase in signal intensity above baseline following the initial decrease in signal intensity (Cho et al., 2002). Such an increase has been shown to be associated to the massive leakage of contrast agent into the interstitial space of PCLs. The rCBV ratio has been also reported significantly lower in their periphery compared to glioblastomas but not with metastases (Wang et al., 2011). This can obviously be attributed to the higher infiltrations of HGGs compared to PCL and metastases.

9.5.3 MRS Contribution in PCLs

The most specific finding for PCNSL on MRS is an increase in lipid and Cho resonances associated with decreased NAA, Cr, and myoinositol levels (Figure 9.9i and j). It has also been reported that a significant increase in the lipid peaks in PCNSL may help differentiate them from HGGs (Harting et al., 2003). Although lipids typically indicate necrosis, this is not true in the case of PCLs. Increased lipid peaks are attributed to the increased turnover of membrane components in transformed lymphoid cells containing mobile lipids (Tang et al., 2011). The peritumoral area of PCLs should also be surveyed. Like HGGs, the peritumoral area of PCLs demonstrates increased Cho/Cr and Lip+Lac/Cr ratios (Figure 9.9e). Moreover Chawla et al. (2010) suggested presence of infiltrative active lymphocytes and macrophages in areas beyond lymphoma boundaries due to the higher Lip+Lac/Cr ratio in the peritumoral area of PCLs when compared with that of GBMs.

Therefore, in the absence of obvious necrotic parts and heterogeneity, increased lipid concentration combined with a markedly elevated Cho/Cr ratio both intra- and peri-tumorally, can provide important metabolic information, which may improve the distinction between PCLs and other brain tumors.

9.6 Intracranial Abscesses

Focus Point

DWI/DTI
- Lower ADC values in the central cavity of abscesses compared to glioblastomas and metastases.
- The high viscosity and cellularity of pus, results in substantially restricted diffusion.

PWI
- Low vascularity and decreased neoangiogenesis characterizes brain abscesses compared to hypervascular high-grade tumors.
- Lower rCBV values for abscesses mimicking HGG or Meta.

MRS
- Presence of amino acids and/or lactate and lipids and/or acenate and sucinate.
- Normal brain metabolites should be absent.

Brain abscesses are focal intracerebral lesions caused by inflammation and collection of infected material, originating either from local infections (e.g., paranasal sinuses, mastoid air cells of the temporal bone, etc.) or remote infectious sources (e.g., lung, heart, kidney, etc.), within the brain tissue. Those pathogens will produce areas of focal cerebritis with accumulation of purulent exudates in terms of collagenous substance in the brain tissue, usually surrounded by a well-vascularized capsule (Grigoriadis and Gold, 1997; Chan et al., 2002; Pal et al., 2010). The imaging findings on conventional MRI, are increased signal intensity on T2-weighted images with associated peritumoral edema, increased signal intensity on DW images and ring-shaped contrast enhancement (Figure 9.10). Hence, abscesses may mimic HGGs or solitary metastases with similar imaging features, especially if cystic or necrotic containing pus regions are involved, thus further complicating differential diagnosis (Cha, 2006; Chiang et al., 2009; Hakyemez et al., 2006a).

9.6.1 DWI DTI Contribution in Abscesses

DWI and DTI metrics can be used to differentiate between abscesses and other cystic lesions (Chan et al., 2002; Chang et al., 2002; Nath et al., 2009; Reiche et al., 2010). Abscesses present high signal on diffusion and lower ADC values compared to glioblastomas and metastases, which a can be explained due to the higher viscosity and cellularity of pus, compared to the cystic or necrotic areas of tumors, which results in substantially restricted diffusion (Lai et al., 2002; Reiche et al., 2010; Gupta et al., 2005). Significant differences in diffusion results have been also observed in the periphery of lesions. Chan et al. investigated the capsular wall of abscesses and reported hypointense DW images and higher ADC values compared to the hyperintense tumor wall, associated with lower ADC values. The authors suggested that a possible explanation is the increased extracellular fluid accumulation in the abscess wall due to inflammation versus the closely packed malignant cells in the periphery of tumors

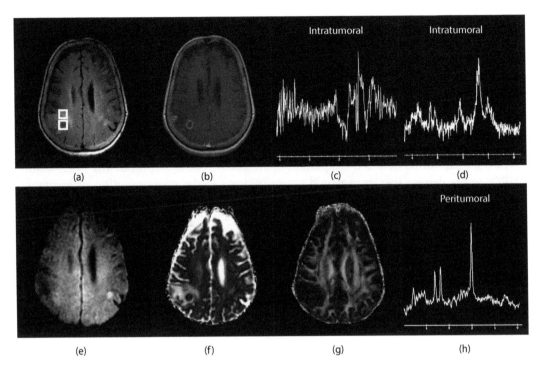

FIGURE 9.10   A case of an intracranial abscesses. (a) T2 FLAIR image of the lesion with associated peritumoral edema voxel's placement, (b) ring-shaped contrast enhancement on a T1-weighted post-contrast image. The lesion is hyperintense on the DW image (e), which results in a hypointense appearance on the ADC (f), and FA (g) maps. The corresponding intratumoral (c and d) and peritumoral (h) spectra are also depicted.

(Chan et al., 2002). Furthermore, high fractional anisotropy in the cavity of brain abscesses versus other cystic intracranial lesions has been reported by Toh et al. (2011). Interestingly they have reported FA values as high as the normal white matter, apparently suggesting an oriented structure within the abscess cavity. This has been attributed to a possible oriented and organized structure of inflammatory cells, owing to cell adhesion. Although these findings can be helpful, they are of course not absolute and should be considered with caution.

9.6.2 Perfusion Contribution in Abscesses

rCBV measurements can have a significant contribution in the differentiation of nonneoplastic processes such as infectious abscesses versus cystic brain tumor such as GBMs and metastases. The main differentiation comes from the lower rCBV ratios in the abscess wall, compared to the periphery of other rim-enhancing lesions (Chiang et al., 2009; Holmes et al., 2004). This is attributed to the higher vascularity and increased neo-angiogenesis of high-grade tumors compared to intracranial abscesses. Hence, peripheral rCBV less than or equal to that of the surrounding WM should suggest the possibility of abscess and it seems that the capillary density differences may enable their distinct discrimination.

Nonetheless, comparisons of vascular permeability between pyogenic abscesses and HGGs and/or metastases has been performed with dynamic contrast-enhanced (DCE) MRI based on the quantification of the volume transfer constant K^{trans}, which also appears to be useful in differentiating infective from neoplastic brain lesions (Haris et al., 2008).

9.6.3 MRS Contribution in Abscesses

MRS can noninvasively contribute to the differential diagnosis between abscesses and HGGs or metastases. All cystic tumors and abscesses appear with increased lactate, which is a nonspecific metabolite that results from anaerobic glycolysis (see Figure 9.10). The typical spectrum of an abscess cavity may appear significantly different from that of necrotic or cystic brain tumor. The main findings are the resonances of amino acids (valine, leucine, and isoleucine) (0.9 ppm), acetate (1.9 ppm), alanine (1.5 ppm), and lactate (1.3 ppm) (Kadota et al., 2001). On the other hand, cerebral abscesses contain no neurons, therefore NAA and Cr/PCr should not be detected (Kapsalaki et al., 2008). The detection of any NAA and/or Cr/PCr in an abscess spectrum would either indicate signal contamination or erroneous interpretation of acetate peak as NAA (Kadota et al., 2001). Similarly, no tCho peak should be present in an abscesses spectrum because there are no membranous structures in its necrotic core (Lai et al., 2005).

The increased levels of lactate, acetate, and succinate presumably originate from the infecting microorganisms, fermentation and enhanced glycolysis. Amino acids such as valine and leucine are known to be the end products of proteolysis by enzymes released by neutrophils in pus (Mader et al., 2008). It should be noted however, that abscesses of tuberculous origin can be characterized by the predominant presence of lipids and moderate increase of tCho (Pretell et al., 2005).

Last, although the presence of amino acids can be considered a sensitive marker of pyogenic abscesses, its absence cannot rule out an abscess of pyogenic origin (Pal et al., 2010). Once again, the peritumoral hypothesis might give an aid towards differential diagnosis. Lai et al. (2008) found increased Cho/Cr and Cho/NAA ratios at the enhancing rim of GBMs, but not for abscesses (as in Figure 9.10h). In the rim of the abscesses only a mild elevation of Cho/Cr was observed, which was attributed to the diminished Cr levels due to general breakdown of the energy metabolism, and not to the increase of Cho (Lai et al., 2005). Thus, the rim enhancing portion may be also advantageous to distinguish between HGGs and abscesses although a metastasis cannot be ruled out.

9.7 Summary and Conclusion

Accurate diagnosis, classification and grading of cerebral lesions are the primary concerns in neuro-oncological imaging. Over the last decades, there has been a rapid evolution in the detection of structural abnormalities, localization and assessment of the extent of the lesion, with advanced MR techniques providing additional insight to tumors' physiology, such as microarchitecture, microvasculature and cellular biochemistry. It is now clear that a multi-parametric analysis substantially improves diagnostic accuracies over conventional MRI alone, and highlights the underlying pathophysiology.

In an attempt to lead the reader of this book to a more comprehensive understanding of the aforementioned mechanisms and techniques, as well as to summarize the correlation and the potential diagnostic outcome of the metrics emerging from the advanced techniques discussed, Table 9.2 includes an outline of all the findings regarding the different types of cerebral pathology. Nevertheless, it must be noted that these are quite complicated relationships; hence, please keep in mind that they are indicative and can be altered, under certain conditions.

TABLE 9.2 Summary of the Correlation of T2/DWI Signals, ADC/FA, rCBV Measurements, and MRS Metabolite Ratios (Intratumoral and Peritumoral), Regarding Different Types of Cerebral Pathology

Pathology	T2	DWI	ADC	FA	rCBV	K^{trans}	Cho/Cr	Lip/Cr	Other Metabol	Cho/Cr‡	Lip/Cr‡
Abscess	˅	↑	↓	↑	-˅	↑	↓	↑	Suc/Acc	–	–
Oligodendroglioma	˅	˅	˅	↓	↑†	-˅	-˅	–	^Myo	–	–
Low-Grade Astrocytoma	˅	↑	↑	↓	-˅	-˅	-˅	–	^Myo	–	–
Anaplastic Astrocytoma	↑	↑	˅	↓	˅	˅	↑	˅	↓NAA	˅	–
Meningioma	↑	↑	↑	˅	Very high	↑	–	–	Ala/Glx, no NAA	–	–
Meningioma (atypical)	↑	↑	↓	↑*˅	↑	↑	↑	↑	Ala/Glx, no NAA	–	–
GBM	↑	˅	˅	↓	Very high	↑	↑	↑	–	↑	˅
Metastasis	↑	˅	˅	↓	↑	↑	↑	↑	–	–	–
Lymphoma	↑	Very high	˅	↓**	↓*	˅	↑	↑	–	˅	˅
Vasogenic Edema	↑	–	˅	˅**	˅	˅	↑	˅	–	–	–
Necrotic tumor	↑	↓	↑	↓	–	–	↑	↑	–	↑	˅

‡Denotes peritumoral region, *Compared to high-grade tumors, **Contradicting results, †Rule exception ↓Decrease, ↑Increase, ^Moderate increase, ˅Moderate decrease, – Isointense/no change/normal

References

Altman, D. A., Atkinson, D. S., and Brat, D. J. (2007). Best cases from the AFIP: Glioblastoma multiforme. *Radiographics: A Review Publication of the Radiological Society of North America, Inc, 27*(3), 883–888.

Bataille, B., Delwail, V., Guy, G., Ingrand, P., Lapierre, F., Menet, E. et al. (2000). Primary intracerebral malignant lymphoma: Report of 248 cases. *Journal of Neurosurgery, 92*(2), 261–266.

Batjer, H. H., and Loftus, C. M. (2003). *Textbook of Neurological Surgery: Principles and Practice: Volume two* (chap. 102, pp. 1257–1270). Philadelphia, PA: Lippincott, Williams & Wilkins.

Bauer, A. H., Erly, W., Moser, F. G., Maya, M., and Nael, K. (2015). Differentiation of solitary brain metastasis from glioblastoma multiforme: A predictive multiparametric approach using combined MR diffusion and perfusion. *Neuroradiology, 57*(7), 697–703. doi:10.1007/s00234-015-1524-6

Bendini, M., Marton, E., Feletti, A., Rossi, S., Curtolo, S., Inches, I. et al. (2011). Primary and metastatic intraaxial brain tumors: Prospective comparison of multivoxel 2D chemical-shift imaging (CSI) proton MR spectroscopy, perfusion MRI, and histopathological findings in a group of 159 patients. *Acta Neurochirurgica, 153*(2), 403–412. doi:10.1007/s00701-010-0833-0

Berger, M. S. (2003). Perfusion MR and the evaluation of meningiomas: Is it important surgically? *AJNR American Journal of Neuroradiology, 24*(8), 1499–1500.

Bertossi, M., Virgintino, D., Maiorano, E., Occhiogrosso, M., and Roncali, L. (1997). Ultrastructural and morphometric investigation of human brain capillaries in normal and peritumoral tissues. *Ultrastructural Pathology, 21*(1), 41–49. doi:10.3109/01913129709023246

Brynolfsson, P., Nilsson, D., Henriksson, R., Hauksson, J., Karlsson, M., Garpebring, A. et al. (2014). ADC texture-An imaging biomarker for high-grade glioma? *Medical Physics, 41*(10), 101903. doi:10.1118/1.4894812

Buetow, M. P., Buetow, P. C., and Smirniotopoulos, J. G. (1991). Typical, atypical, and misleading features in meningioma. *Radiographics: A Review Publication of the Radiological Society of North America, Inc, 11*(6), 1087–1106.

Buhl, R., Nabavi, A., Wolff, S., Hugo, H. H., Alfke, K., Jansen, O., and Mehdorn, H. M. (2007). MR spectroscopy in patients with intracranial meningiomas. *Neurological Research, 29*(1), 43–46. doi:10.117 9/174313206x153824

Bulakbasi, N., Kocaoglu, M., Ors, F., Tayfun, C., and Uçöz, T. (2003). Combination of single-voxel proton MR spectroscopy and apparent diffusion coefficient calculation in the evaluation of common brain tumors. AJNR. *American Journal of Neuroradiology, 24*(2), 225–233.

Calli, C., Kitis, O., Yunten, N., Yurtseven, T., Islekel, S., and Akalin, T. (2006). Perfusion and diffusion MR imaging in enhancing malignant cerebral tumors. *European Journal of Radiology, 58*(3), 394–403. doi:10.1016/j.ejrad.2005.12.032

Castillo, M., and Kwock, L. (1999). Clinical applications of proton magnetic resonance spectroscopy in the evaluation of common intracranial tumors. *Topics in Magnetic Resonance Imaging, 10*(2), 104–113. doi:10.1097/00002142-199904000-00003

Castillo, M., Kwock, L., and Smith, J. K. (2000). Correlation of myo-inositol levels and grading of cerebral astrocytomas. AJNR. *American Journal of Neuroradiology, 21*(9), 1645–1649.

Cha, S. (2004). Perfusion MR imaging of brain tumors. *Topics in Magnetic Resonance Imaging: TMRI, 15*(5), 279–289.

Cha, S. (2006). Update on brain tumor imaging: From anatomy to physiology. AJNR. *American Journal of Neuroradiology, 27*(3), 475–487.

Cha, S., Knopp, E. A., Johnson, G., Wetzel, S. G., Litt, A. W., and Zagzag, D. (2002). Intracranial mass lesions: Dynamic contrast-enhanced susceptibility-weighted echo-planar perfusion MR imaging. *Radiology, 223*(1), 11–29.

Cha, S., Tihan, T., Crawford, F., Fischbein, N. J., Chang, S., Bollen, A. et al. (2005). Differentiation of low-grade oligodendrogliomas from low-grade astrocytomas by using quantitative blood-volume measurements derived from dynamic susceptibility contrast-enhanced MR imaging. *AJNR American Journal of Neuroradiology, 26*(2), 266–273.

Chan, J. H., Tsui, E. Y., Chau, L. F., Chow, K. Y., Chan, M. S., Yuen, M. K. et al. (2002). Discrimination of an infected brain tumor from a cerebral abscess by combined MR perfusion and diffusion imaging. *Computerized Medical Imaging and Graphics, 26*(1), 19–23. doi:10.1016/s0895-6111(01)00023-4

Chang, S. C., Lai, P. H., Chen, W. L., Weng, H. H., Ho, J. T., Wang, J. S. et al. (2002). Diffusion-weighted MRI features of brain abscess and cystic or necrotic brain tumors. *Clinical Imaging, 26*(4), 227–236. doi:10.1016/s0899-7071(02)00436-9

Chawla, S., Zhang, Y., Wang, S., Chaudhary, S., Chou, C., Orourke, D. M. et al. (2010). Proton magnetic resonance spectroscopy in differentiating glioblastomas from primary cerebral lymphomas and brain metastases. *Journal of Computer Assisted Tomography, 34*(6), 836–841. doi:10.1097/rct.0b013e3181ec554e

Chen, Y., Shi, Y., and Song, Z. (2010). Differences in the architecture of low-grade and high-grade gliomas evaluated using fiber density index and fractional anisotropy. *Journal of Clinical Neuroscience, 17*(7), 824–829. doi:10.1016/j.jocn.2009.11.022

Chiang, I. C., Kuo, Y. T., Lu, C. Y., Yeung, K. W., Lin, W. C., Sheu, F. O., and Liu, G. C. (2004). Distinction between high-grade gliomas and solitary metastases using peritumoral 3-T magnetic resonance spectroscopy, diffusion, and perfusion imagings. *Neuroradiology, 46*(8), 619–627. doi:10.1007/s00234-004-1246-7

Chiang, I. C., Chiu, M. L., Hsieh, T. J., Kuo, Y. T., Liu, G. C., and Lin, W. C. (2009). Distinction between pyogenic brain abscess and necrotic brain tumour using 3-tesla MR spectroscopy, diffusion and perfusion imaging. *The British Journal of Radiology, 82*(982), 813–820.

Cho, S. K., Byun, H. S., Kim, J. H., Moon, C. H., Na, D. G., Ryoo, J. W., and Roh, H. G. (2002). Perfusion MR imaging: Clinical utility for the differential diagnosis of various brain tumors. *Korean Journal of Radiology, 3*(3), 171–179.

Choi, H. S., Kim, A. H., Ahn, S. S., Shin, N. Y., Kim, J., and Lee, S. K. (2013). Glioma grading capability: Comparisons among parameters from dynamic contrast-enhanced MRI and ADC value on DWI. *Korean Journal of Radiology, 14*(3), 487–492. doi:10.3348/kjr.2013.14.3.487

Collins, J. M., and Christoforidis, G. A. (2016). Meningeal tumors. In H. B. Newton (Ed.), *Handbook of Neuro-Oncology. Neuroimaging* (pp. 519–542). Amsterdam: Academic Press.

De Belder, F. E., Oot, A. R., Hecke, W. V., Venstermans, C., Menovsky, T., Marck, V. V. et al. (2012). Diffusion tensor imaging provides an insight into the microstructure of meningiomas, high-grade gliomas, and peritumoral edema. *Journal of Computer Assisted Tomography, 36*(5), 577–582. doi:10.1097/rct.0b013e318261e913

Devos, A., Simonetti, A., Graaf, M. V., Lukas, L., Suykens, J., Vanhamme, L. et al. (2005). The use of multivariate MR imaging intensities versus metabolic data from MR spectroscopic imaging for brain tumour classification. *Journal of Magnetic Resonance, 173*(2), 218–228. doi:10.1016/j.jmr.2004.12.007

Di Costanzo, A. D., Armillotta, M., Canalls, L., Carriero, A., Giannatempo, G. M., Maggialetti, A. et al. (2008). Role of perfusion-weighted imaging at 3 Tesla in the assessment of malignancy of cerebral gliomas. *La Radiologia Medica, 113*(1), 134–143.

Ellison, D., Love, S., Chimelli, L., Harding, B., Lowe, J. S., Vinters, H. V. et al. (2004). Meningiomas. *In Neuropathology: A Reference Text of CNS Pathology* (pp. 703–716). Edinburgh, UK: Mosby.

Fan, G. G., Deng, Q. L., Wu, Z. H., and Guo, Q. Y. (2006). Usefulness of diffusion/perfusion-weighted MRI in patients with non-enhancing supratentorial brain gliomas: A valuable tool to predict tumour grading? *The British Journal of Radiology, 79*(944), 652–658. doi:10.1259/bjr/25349497

Ferda, J., Choc, M., Ferdová, E., Horemuzová, J., Kastner, J., Kreuzberg, B., and Mukensnabl, P. (2010). Diffusion tensor magnetic resonance imaging of glial brain tumors. *European Journal of Radiology, 74*(3), 428–436.

Filippi, C. G., Edgar, M. A., Heier, L. A., Prowda, J. C., Uluğ, A. M., and Zimmerman, R. D. (2001). Appearance of meningiomas on diffusion-weighted images: Correlating diffusion constants with histopathologic findings. *AJNR American Journal of Neuroradiology, 22*(1), 65–72.

Fountas, K. N., Gotsis, S. D., Johnston, K. W., Kapsalaki, E. Z., Kapsalakis, J. Z., Papadakis, N. et al. (2000). In vivo proton magnetic resonance spectroscopy of brain tumors. *Stereotactic and Functional Neurosurgery, 74*(2), 83–94.

Georgiadis, P., Kostopoulos, S., Cavouras, D., Glotsos, D., Kalatzis, I., Sifaki, K. et al. (2011). Quantitative combination of volumetric MR imaging and MR spectroscopy data for the discrimination of meningiomas from metastatic brain tumors by means of pattern recognition. *Magnetic Resonance Imaging*, *29*(4), 525–535. doi:10.1016/j.mri.2010.11.006

Gillard, J. H., Waldman, A. D., and Barker, P. B. (2005). *Clinical MR Neuroimaging: Physiological and Functional Techniques*. Cambridge, UK: Cambridge University Press.

Ginat, D. T., Mangla, R., Wang, H. Z., and Yeaney, G. (2010). Correlation of diffusion and perfusion MRI with Ki-67 in high-grade meningiomas. *AJR American Journal of Roentgenology*, *195*(6), 1391–1395.

Go, J. L., Kim, P. E., and Lee, S. C. (2006). Imaging of primary central nervous system lymphoma. *Neurosurgical Focus*, *21*(5), E4.

Goebell, E., Paustenbach, S., Vaeterlein, O., Ding, X., Heese, O., Fiehler, J. et al. (2006). Low-grade and anaplastic gliomas: Differences in architecture evaluated with diffusion-tensor MR imaging. *Radiology*, *239*(1), 217–222. doi:10.1148/radiol.2383050059

Grigoriadis, E. and Gold, W. L. (1997). Pyogenic brain abscess caused by Streptococcus pneumoniae: Case report and review. *Clinical infectious Diseases: An official publication of the Infectious Diseases Society of America*, *25*(5), 1108–1112.

Guo, A. C., Cummings, T. J., Dash, R. C., and Provenzale, J. M. (2002). Lymphomas and high-grade astrocytomas: Comparison of water diffusibility and histologic characteristics. *Radiology*, *224*(1), 177–183. doi:10.1148/radiol.2241010637

Gupta, R. K., Hasan, K. M., Husain, M., Jha, D., Mishra, A. M., Narayana, P. A., and Prasad, K. N. (2005). High fractional anisotropy in brain abscesses versus other cystic intracranial lesions. *AJNR American Journal of Neuroradiology*, *26*(5), 1107–1114.

Hakyemez, B., Atahan, S., Erdogan, C., Ercan, I., Ergin, N., and Uysal, S. (2005). High-grade and low-grade gliomas: Differentiation by using perfusion MR imaging. *Clinical Radiology*, *60*(4), 493–502.

Hakyemez, B., Dusak, A., Erdogan, C., Gokalp, G., and Parlak, M. (2010). Solitary metastases and high-grade gliomas: Radiological differentiation by morphometric analysis and perfusion-weighted MRI. *Clinical Radiology*, *65*(1), 15–20.

Hakyemez, B., Bolca, N., Erdogan, C., Gokalp, G., Parlak, M., and Yildirim, N. (2006a). Evaluation of different cerebral mass lesions by perfusion-weighted MR imaging. *Journal of Magnetic Resonance Imaging: JMRI*, *24*(4), 817–824.

Hakyemez, B., Yildirim, N., Erdoðan, C., Kocaeli, H., Korfali, E., and Parlak, M. (2006b). Meningiomas with conventional MRI findings resembling intraaxial tumors: Can perfusion-weighted MRI be helpful in differentiation? *Neuroradiology*, *48*(10), 695–702. doi:10.1007/s00234-006-0115-y

Hakyemez, B., Yıldırım, N., Gokalp, G., Erdogan, C., and Parlak, M. (2006c). The contribution of diffusion-weighted MR imaging to distinguishing typical from atypical meningiomas. *Neuroradiology*, *48*(8), 513–520. doi:10.1007/s00234-006-0094-z

Haris, M., Gupta, R. K., Singh, A., Husain, N., Husain, M., Pandey, C. M. et al. (2008). Differentiation of infective from neoplastic brain lesions by dynamic contrast-enhanced MRI. *Neuroradiology*, *50*(6), 531–540. doi:10.1007/s00234-008-0378-6

Harting, I., Hartmann, M., Jost, G., Sommer, C., Ahmadi, R., Heiland, S., and Sartor, K. (2003). Differentiating primary central nervous system lymphoma from glioma in humans using localised proton magnetic resonance spectroscopy. *Neuroscience Letters*, *342*(3), 163–166. doi:10.1016/s0304-3940(03)00272-6

Hartmann, M., Heiland, S., Harting, I., Ludwig, R., Sommer, C., Sartor, K., and Tronnier, V. M. (2003). Distinguishing of primary cerebral lymphoma from high-grade glioma with perfusion-weighted magnetic resonance imaging. *Neuroscience Letters*, *338*(2), 119–122.

Hattingen, E., Raab, P., Franz, K., Zanella, F. E., Lanfermann, H., and Pilatus, U. (2008). Myo-Inositol: A marker of reactive astrogliosis in glial tumors? *NMR in Biomedicine*, *21*(3), 233–241. doi:10.1002/nbm.1186

Helenius, J., Aronen, H. J., Carano, R. A., Kangasmäki, A., Kaste, M., Perkiö, J. et al. (2002). Diffusion-weighted MR imaging in normal human brains in various age groups. *AJNR American Journal of Neuroradiology, 23*(2), 194–199.

Holmes, T. M., Petrella, J. R., and Provenzale, J. M. (2004). Distinction between cerebral abscesses and high-grade neoplasms by dynamic susceptibility contrast perfusion MRI. *American Journal of Roentgenology, 183*(5), 1247–1252. doi:10.2214/ajr.183.5.1831247

Howe, F. A., Barton, S. J., Bell, B. A., Cudlip, S. A., Doyle, V. L., Griffiths, J. R. et al. (2003). Metabolic profiles of human brain tumors using quantitative in vivo 1H magnetic resonance spectroscopy. *Magnetic Resonance in Medicine, 49*(2), 223–232.

Hu, X., Wong, K. K., Young, G. S., Guo, L., and Wong, S. T. (2011). Support vector machine multiparametric MRI identification of pseudoprogression from tumor recurrence in patients with resected glioblastoma. *Journal of Magnetic Resonance Imaging, 33*(2), 296–305. doi:10.1002/jmri.22432

Inoue, T., Ogasawara, K., Beppu, T., Ogawa, A., and Kabasawa, H. (2005). Diffusion tensor imaging for preoperative evaluation of tumor grade in gliomas. *Clinical Neurology and Neurosurgery, 107*(3), 174–180. doi:10.1016/j.clineuro.2004.06.011

Ishimaru, H., Morikawa, M., Iwanaga, S., Kaminogo, M., Ochi, M., and Hayashi, K. (2001). Differentiation between high-grade glioma and metastatic brain tumor using single-voxel proton MR spectroscopy. *European Radiology, 11*(9), 1784–1791. doi:10.1007/s003300000814

Jung, B. C., Arevalo-Perez, J., Lyo, J. K., Holodny, A. I., Karimi, S., Young, R. J., and Peck, K. K. (2015). Comparison of glioblastomas and brain metastases using dynamic contrast-enhanced perfusion MRI. *Journal of Neuroimaging, 26*(2), 240–246. doi:10.1111/jon.12281

Kadota, O., Kohno, K., Ohue, S., Kumon, Y., Sakaki, S., Kikuchi, K., and Miki, H. (2001). Discrimination of brain abscess and cystic tumor by in vivo proton magnetic resonance spectroscopy. *Neurologia Medico-Chirurgica, 41*(3), 121–126. doi:10.2176/nmc.41.121

Kapsalaki, E. Z., Fountas, K. N., and Gotsis, E. D. (2008). The role of proton magnetic resonance spectroscopy in the diagnosis and categorization of cerebral abscesses. *Neurosurgical Focus, 24*(6), E7.

Kimura, T., Sako, K., Gotoh, T., Tanaka, K., and Tanaka, T. (2001). In vivo single-voxel proton MR spectroscopy in brain lesions with ring-like enhancement. *NMR in Biomedicine, 14*(6), 339–349. doi:10.1002/nbm.711.abs

Kono, K., Inoue, Y., Morino, M., Nakayama, K., Ohata, K., Shakudo, M. et al. (2001). The role of diffusion-weighted imaging in patients with brain tumors. AJNR. *American Journal of Neuroradiology, 22*(6), 1081–1088.

Kousi, E., Tsougos, I., Fountas, K., Theodorou, K., Tsolaki, E., Fezoulidis, I., and Kapsalaki, E. (2012). Distinct peak at 3.8 ppm observed by 3T MR spectroscopy in meningiomas, while nearly absent in high-grade gliomas and cerebral metastases. *Molecular Medicine Reports, 5*(4), 1011–1018.

Kremer, S., Grand, S., Rémy, C., Pasquier, B., Benabid, A. L., Bracard, S., and Bas, J. F. (2004). Contribution of dynamic contrast MR imaging to the differentiation between dural metastasis and meningioma. *Neuroradiology, 46*(8), 642–648. doi:10.1007/s00234-004-1194-2

Lai, P. H., Chen, W. L., Ho, J. T., Hsu, S. S., Pan, H. B., Wang, J. S., and Yang, C. F. (2002). Brain abscess and necrotic brain tumor: Discrimination with proton MR spectroscopy and diffusion-weighted imaging. *AJNR American Journal of Neuroradiology, 23*(8), 1369–1377.

Lai, P. H., Ding, S., Hsu, S. S., Hsiao, C. C., Li, K. T., Pan, H. B. et al. (2005). Pyogenic brain abscess: Findings from in vivo 1.5-T and 11.7-T in vitro proton MR spectroscopy. *AJNR American Journal of Neuroradiology, 26*(2), 279–288.

Lam, W., Poon, W., and Metreweli, C. (2002). Diffusion MR imaging in glioma: Does it have any role in the pre-operation determination of grading of glioma? *Clinical Radiology, 57*(3), 219–225. doi:10.1053/crad.2001.0741

Law, M., Arnett, J., Cha, S., Johnson, G., Knopp, E. A., and Litt, A. W. (2002). High-grade gliomas and solitary metastases: Differentiation by using perfusion and proton spectroscopic MR imaging. *Radiology, 222*(3), 715–721.

Law, M., Babb, J. S., Cha, S., Johnson, G., Knopp, E. A., Wang, H. et al. (2003). Glioma grading: Sensitivity, specificity, and predictive values of perfusion MR imaging and proton MR spectroscopic imaging compared with conventional MR imaging. *AJNR American Journal of Neuroradiology, 24*(10), 1989–1998.

Law, M., Babb, J. S., Chheang, S., Gruber, M. L., Golfinos, J. G., Johnson, G., . . . Zagzag, D. (2008). Gliomas: Predicting time to progression or survival with cerebral blood volume measurements at dynamic susceptibility-weighted contrast-enhanced perfusion MR imaging. *Radiology, 247*(2), 490–498.

Law, M., Young, R., Babb, J., Rad, M., Sasaki, T., Zagzag, D., and Johnson, G. (2006). Comparing perfusion metrics obtained from a single compartment versus pharmacokinetic modeling methods using dynamic susceptibility contrast-enhanced perfusion MR imaging with glioma grade. *AJNR American Journal of Neuroradiology, 27*(9), 1975–1982.

Lee, S. (2012). Diffusion tensor and perfusion imaging of brain tumors in high-field MR imaging. *Neuroimaging Clinics of North America, 22*(2), 123–134, ix.

Lee, H. Y., Na, D. G., Song, I., Lee, D. H., Seo, H. S., Kim, J., and Chang, K. (2008). Diffusion-tensor imaging for glioma grading at 3-T magnetic resonance imaging. *Journal of Computer Assisted Tomography, 32*(2), 298–303. doi:10.1097/rct.0b013e318076b44d

Lee, I. H., Kim, S. T., Kim, H., Kim, K. H., Jeon, P., and Byun, H. S. (2010). Analysis of perfusion weighted image of CNS lymphoma. *European Journal of Radiology, 76*(1), 48–51. doi:10.1016/j.ejrad.2009.05.013

Lev, M. H., Barest, G. D., Chiocca, E. A., Csavoy, A. N., Fitzek, M. M., Gonzalez, R. G. et al. (2004). Glial tumor grading and outcome prediction using dynamic spin-echo MR susceptibility mapping compared with conventional contrast-enhanced MR: Confounding effect of elevated rCBV of oligodendrogliomas [corrected]. *AJNR American Journal of Neuroradiology, 25*(2), 214–221.

Li, X., Zhu, Y., Kang, H., Zhang, Y., Liang, H., Wang, S., and Zhang, W. (2015). Glioma grading by microvascular permeability parameters derived from dynamic contrast-enhanced MRI and intratumoral susceptibility signal on susceptibility weighted imaging. *Cancer Imaging, 15*(1), 4. doi:10.1186/s40644-015-0039-z

Liu, X., Tian, W., Kolar, B., Yeaney, G. A., Qiu, X., Johnson, M. D., and Ekholm, S. (2011). MR diffusion tensor and perfusion-weighted imaging in preoperative grading of supratentorial nonenhancing gliomas. *Neuro-Oncology, 13*(4), 447–455. doi:10.1093/neuonc/noq197

Lu, S., Ahn, D., Johnson, G., Law, M., Zagzag, D., and Grossman, R. I. (2004). Diffusion-tensor MR imaging of intracranial neoplasia and associated peritumoral edema: Introduction of the tumor infiltration index. *Radiology, 232*(1), 221–228. doi:10.1148/radiol.2321030653.

Mader, I., Rauer, S., Gall, P., and Klose, U. (2008). ¹H MR spectroscopy of inflammation, infection and ischemia of the brain. *European Journal of Radiology, 67*(2), 250–257. doi:10.1016/j.ejrad.2008.02.033

Maia, A. C., Ferraz, F. A., Gabbai, A. A., Malheiros, S. M., Rocha, A. J., Silva, C. J., and Stávale, J. N. (2005). MR cerebral blood volume maps correlated with vascular endothelial growth factor expression and tumor grade in nonenhancing gliomas. AJNR. *American Journal of Neuroradiology, 26*(4), 777–783.

Maier, S. E., Gudbjartsson, H., Patz, S., Hsu, L., Lovblad, K. O., Edelman, R. R. et al. (1998). Line scan diffusion imaging: Characterization in healthy subjects and stroke patients. *American Journal of Roentgenology, 171*(1), 85–93. doi:10.2214/ajr.171.1.9648769

Maier, S. E., Sun, Y., and Mulkern, R. V. (2010). Diffusion imaging of brain tumors. *NMR in Biomedicine, 23*(7), 849–864. doi:10.1002/nbm.1544

Majós, C., Alonso, J., Aguilera, C., Acebes, J. J., Arús, C., Gili, J. et al. (2004). Brain tumor classification by proton MR spectroscopy: Comparison of diagnostic accuracy at short and long TE. AJNR. *American Journal of Neuroradiology, 25*(10), 1696–1704.

Möller-Hartmann, W., Herminghaus, S., Krings, T., Lanfermann, H., Marquardt, G., Pilatus, U., and Zanella, F. E. (2002). Clinical application of proton magnetic resonance spectroscopy in the diagnosis of intracranial mass lesions. *Neuroradiology, 44*(5), 371–381.

Morita, K., Matsuzawa, H., Fujii, Y., Tanaka, R., Kwee, I. L., and Nakada, T. (2005). Diffusion tensor analysis of peritumoral edema using lambda chart analysis indicative of the heterogeneity of the microstructure within edema. *Journal of Neurosurgery, 102*(2), 336–341. doi:10.3171/jns.2005.102.2.0336

Nagar, V., Ye, J., Ng, W., Chan, Y., Hui, F., Lee, C., and Lim, C. (2008). Diffusion-weighted MR imaging: Diagnosing atypical or malignant meningiomas and detecting tumor dedifferentiation. *American Journal of Neuroradiology*, 29(6), 1147–1152. doi:10.3174/ajnr.a0996

Nath, K., Agarwal, M., Ramola, M., Husain, M., Prasad, K. N., Rathore, R. K., and Gupta, R. K. (2009). Role of diffusion tensor imaging metrics and in vivo proton magnetic resonance spectroscopy in the differential diagnosis of cystic intracranial mass lesions. *Magnetic Resonance Imaging*, 27(2), 198–206. doi:10.1016/j.mri.2008.06.006

Opstad, K. S., Murphy, M. M., Wilkins, P. R., Bell, B. A., Griffiths, J. R., and Howe, F. A. (2004). Differentiation of metastases from high-grade gliomas using short echo time ^{1}H spectroscopy. *Journal of Magnetic Resonance Imaging*, 20(2), 187–192. doi:10.1002/jmri.20093

Pal, D., Bhattacharyya, A., Husain, M., Prasad, K., Pandey, C., and Gupta, R. (2010). In vivo proton MR spectroscopy evaluation of pyogenic brain abscesses: A report of 194 cases. *American Journal of Neuroradiology*, 31(2), 360–366. doi:10.3174/ajnr.a1835

Patchell, R. A. (1991). Brain metastases. *Neurologic Clinics*, 9(4), 817–827.

Pauleit, D., Langen, K., Floeth, F., Hautzel, H., Riemenschneider, M. J., Reifenberger, G., and Müller, H. (2004). Can the apparent diffusion coefficient be used as a noninvasive parameter to distinguish tumor tissue from peritumoral tissue in cerebral gliomas? *Journal of Magnetic Resonance Imaging*, 20(5), 758–764. doi:10.1002/jmri.20177

Pavlisa, G., Rados, M., Pavlisa, G., Pavic, L., Potocki, K., and Mayer, D. (2009). The differences of water diffusion between brain tissue infiltrated by tumor and peritumoral vasogenic edema. *Clinical Imaging*, 33(2), 96–101. doi:10.1016/j.clinimag.2008.06.035

Pavlisa, G., Rados, M., Pazanin, L., Padovan, R. S., Ozretic, D., and Pavlisa, G. (2008). Characteristics of typical and atypical meningiomas on ADC maps with respect to schwannomas. *Clinical Imaging*, 32(1), 22–27. doi:10.1016/j.clinimag.2007.07.007

Perry, A., Louis, D. N., Scheithauer, B. W., et al. (2007). Meningiomas. In D. N. Louis, H. Ohgaki, O. D. Wiestler, and W. K. Cavenee (Eds.), *WHO Classification of Tumours of the Central Nervous System* (pp. 164–172). Lyon, France: IARC Press.

Porto, L., Kieslich, M., Franz, K., Lehrnbecher, T., Zanella, F., Pilatus, U., and Hattingen, E. (2011). MR spectroscopy differentiation between high and low grade astrocytomas: A comparison between paediatric and adult tumours. *European Journal of Paediatric Neurology*, 15(3), 214–221. doi:10.1016/j.ejpn.2010.11.003

Pretell, E. J., Martinot, C., Garcia, H. H., Alvarado, M., Bustos, J. A., and Martinot, C. (2005). Differential diagnosis between cerebral tuberculosis and neurocysticercosis by magnetic resonance spectroscopy. *Journal of Computer Assisted Tomography*, 29(1), 112–114. doi:10.1097/01.rct.0000149959.63294.8f

Price, S. J. (2010). Advances in imaging low-grade gliomas. *Advances and Technical Standards in Neurosurgery*, 35, 1–34. doi:10.1007/978-3-211-99481-8_1

Provenzale, J. M., York, G., Moya, M. G., Parks, L., Choma, M., Kealey, S. et al. (2006). Correlation of relative permeability and relative cerebral blood volume in high-grade cerebral neoplasms. *American Journal of Roentgenology*, 187(4), 1036–1042. doi:10.2214/ajr.04.0676

Qi, Z., Chen, X., Geng, D., Li, Y., Li, K., Shen, T., and Wang, Y. (2008). Lipid signal in evaluation of intracranial meningiomas. *Chinese Medical Journal*, 121(23), 2415–2419.

Reiche, W., Schuchardt, V., Hagen, T., Il'Yasov, K. A., Billmann, P., and Weber, J. (2010). Differential diagnosis of intracranial ring enhancing cystic mass lesions—Role of diffusion-weighted imaging (DWI) and diffusion-tensor imaging (DTI). *Clinical Neurology and Neurosurgery*, 112(3), 218–225. doi:10.1016/j.clineuro.2009.11.016

Rijpkema, M., Schuuring, J., Meulen, Y. V., Graaf, M. V., Bernsen, H., Boerman, R., and Heerschap, A. (2003). Characterization of oligodendrogliomas using short echo time ^{1}H MR spectroscopic imaging. *NMR in Biomedicine*, 16(1), 12–18. doi:10.1002/nbm.807

Rizzo, L., Crasto, S. G., Moruno, P. G., Cassoni, P., Rudà, R., Boccaletti, R. et al. (2009). Role of diffusion- and perfusion-weighted MR imaging for brain tumour characterization. *La Radiologia Medica*, 114(4), 645–659.

Ryu, Y. J., Choi, S. H., Park, S. J., Yun, T. J., Kim, J., and Sohn, C. (2014). Glioma: Application of whole-tumor texture analysis of diffusion-weighted imaging for the evaluation of tumor heterogeneity. *PLOS One*, *9*(9), e108335. doi:10.1371/journal.pone.0108335

Santelli, L., Ramondo, G., Puppa, A. D., Ermani, M., Scienza, R., D'Avella, D., and Manara, R. (2010). Diffusion-weighted imaging does not predict histological grading in meningiomas. *Acta Neurochirurgica*, *152*(8), 1315–1319. doi:10.1007/s00701-010-0657-y

Sawaya, R. (2001). Considerations in the diagnosis and management of brain metastases. *Oncology*, *15*(9), 1144–1154, 1157–1158; discussion 1158, 1163–1165.

Scarabino, T., Popolizio, T., Trojsi, F., Giannatempo, G., Pollice, S., Maggialetti, N., and Salvolini, U. (2009). Role of advanced MR imaging modalities in diagnosing cerebral gliomas. *La Radiologia Medica*, *114*(3), 448–460. doi:10.1007/s11547-008-0351-9

Sentürk, S., Cila, A., and Oğuz, K. K. (2009). Dynamic contrast-enhanced susceptibility-weighted perfusion imaging of intracranial tumors: A study using a 3T MR scanner. *Diagnostic and Interventional Radiology*, *15*(1), 3–12.

Server, A., Gadmar, Ø., Haakonsen, M., Josefsen, R., Kulle, B., Kumar, T. et al. (2010). Proton magnetic resonance spectroscopy in the distinction of high-grade cerebral gliomas from single metastatic brain tumors. *Acta Radiologica*, *51*(3), 316–325.

Server, A., Graff, B. A., Josefsen, R., Nordhøy, W., Nakstad, P. H., Orheim, T. E., and Schellhorn, T. (2014). Analysis of diffusion tensor imaging metrics for gliomas grading at 3 T. *European Journal of Radiology*, *83*(3), e156–e165.

Server A, Kulle B, Gadmar ØB, Josefsen R, Kumar T, Nakstad PH. (2011) Measurements of diagnostic examination performance using quantitative apparent diffusion coefficient and proton MR spectroscopic imaging in the preoperative evaluation of tumor grade in cerebral gliomas. *Eur J Radiol*. Nov;*80*(2):462–70. doi: 10.1016/j.ejrad.

Server, A., Kulle, B., Mæhlen, J., Josefsen, R., Schellhorn, T., Kumar, T., and Nakstad, P. (2009). Quantitative apparent diffusion coefficients in the characterization of brain tumors and associated peritumoral edema. *Acta Radiologica*, *50*(6), 682–689. doi:10.1080/02841850902933123

Sibtain, N., Howe, F., and Saunders, D. (2007). The clinical value of proton magnetic resonance spectroscopy in adult brain tumours. *Clinical Radiology*, *62*(2), 109–119. doi:10.1016/j.crad.2006.09.012

Soares, D. P. and Law M. (2009). Magnetic resonance spectroscopy of the brain: Review of metabolites and clinical applications. *Clinical Radiology*, *64*(1), 12–21. doi: 10.1016/j.crad.2008.07.002. Epub 2008 Aug 30.

Stadlbauer, A., Ganslandt, O., Buslei, R., Hammen, T., Gruber, S., Moser, E. et al. (2006). Gliomas: Histopathologic evaluation of changes in directionality and magnitude of water diffusion at diffusion-tensor MR imaging. *Radiology*, *240*(3), 803–810. doi:10.1148/radiol.2403050937

Surawicz, T. S. (1999). Descriptive epidemiology of primary brain and CNS tumors: Results from the Central Brain Tumor Registry of the United States, 1990–1994. *Neuro-Oncology*, *1*(1), 14–25. doi:10.1215/15228517-1-1-14

Svolos, P., Fezoulidis, I., Kousi, E., Kapsalaki, E., Kappas, C., Theodorou, K., and Tsougos, I. (2014). The role of diffusion and perfusion weighted imaging in the differential diagnosis of cerebral tumors: A review and future perspectives. *Cancer Imaging: The Official Publication of the International Cancer Imaging Society*, *14*(1), 20.

Svolos, P., Tsolaki, E., Kapsalaki, E., Theodorou, K., Fountas, K., Fezoulidis, I., and Tsougos, I. (2013). Investigating brain tumor differentiation with diffusion and perfusion metrics at 3T MRI using pattern recognition techniques. *Magnetic Resonance Imaging*, *31*(9), 1567–1577. doi:10.1016/j.mri.2013.06.010

Tang, Y. Z., Booth, T. C., Bhogal, P., Malhotra, A., and Wilhelm, T. (2011). Imaging of primary central nervous system lymphoma. *Clinical Radiology*, *66*(8), 768–777.

Tedeschi, G., Lundbom, N., Raman, R., Bonavita, S., Duyn, J. H., Alger, J. R., and Chiro, G. D. (1997). Increased choline signal coinciding with malignant degeneration of cerebral gliomas: A serial proton magnetic resonance spectroscopy imaging study. *Neurosurgical Focus*, *3*(5). doi:10.3171/foc.1997.3.5.4

Toh, C., Castillo, M., Wong, A., Wei, K., Wong, H., Ng, S., and Wan, Y. (2008a). Differentiation between classic and atypical meningiomas with use of diffusion tensor imaging. *American Journal of Neuroradiology*, *29*(9), 1630–1635. doi:10.3174/ajnr.a1170

Toh, C., Castillo, M., Wong, A., Wei, K., Wong, H., Ng, S., and Wan, Y. (2008b). Primary cerebral lymphoma and glioblastoma multiforme: Differences in diffusion characteristics evaluated with diffusion tensor imaging. *American Journal of Neuroradiology*, *29*(3), 471–475. doi:10.3174/ajnr.a0872

Toh, C. H., Wei, K. C., Ng, S. H., Wan, Y. L., Lin, C. P., and Castillo, M. (2011). Differentiation of brain abscesses from necrotic glioblastomas and cystic metastatic brain tumors with diffusion tensor imaging. *AJNR American Journal of Neuroradiology*, *32*(9), 1646–1651.

Tropine, A., Dellani, P. D., Glaser, M., Bohl, J., Plöner, T., Vucurevic, G. et al. (2007). Differentiation of fibroblastic meningiomas from other benign subtypes using diffusion tensor imaging. *Journal of Magnetic Resonance Imaging*, *25*(4), 703–708. doi:10.1002/jmri.20887

Tropine, A., Vucurevic, G., Delani, P., Boor, S., Hopf, N., Bohl, J., and Stoeter, P. (2004). Contribution of diffusion tensor imaging to delineation of gliomas and glioblastomas. *Journal of Magnetic Resonance Imaging*, *20*(6), 905–912. doi:10.1002/jmri.20217

Tsolaki, E., Svolos, P., Kousi, E., Kapsalaki, E., Fountas, K., Theodorou, K., and Tsougos, I. (2013). Automated differentiation of glioblastomas from intracranial metastases using 3T MR spectroscopic and perfusion data. *International Journal of Computer Assisted Radiology and Surgery*, *8*(5), 751–761. doi:10.1007/s11548-012-0808-0

Tsougos, I., Svolos, P., Kousi, E., Fountas, K., Theodorou, K., Fezoulidis, I., and Kapsalaki, E. (2012). Differentiation of glioblastoma multiforme from metastatic brain tumor using proton magnetic resonance spectroscopy, diffusion and perfusion metrics at 3 T. *Cancer Imaging*, *12*(3), 423–436. doi:10.1102/1470-7330.2012.0038

Tsuruda, J. S., Chew, W. M., Moseley, M. E., and Norman, D. (1990). Diffusion-weighted MR imaging of the brain: Value of differentiating between extraaxial cysts and epidermoid tumors. *American Journal of Roentgenology*, *155*(5), 1059–1065. doi:10.2214/ajr.155.5.2120936

Vamvakas, A., Tsougos, I., Arikidis, N., Kapsalaki, E., Fezoulidis, I., and Costaridou, L. (2016). Local curvature analysis for differentiating Glioblastoma multiforme from solitary metastasis. *2016 IEEE International Conference on Imaging Systems and Techniques (IST), Chania, Greece*. doi:10.1109/ist.2016.7738219

Van Westen, D. V., Brockstedt, S., Englund, E., Lätt, J., and Larsson, E. M. (2006). Tumor extension in high-grade gliomas assessed with diffusion magnetic resonance imaging: Values and lesion-to-brain ratios of apparent diffusion coefficient and fractional anisotropy. *Acta Radiologica*, *47*(3), 311–319.

Verma, R., Zacharaki, E. I., Ou, Y., Cai, H., Chawla, S., Lee, S. K., Melhem, E. R., Wolf, R., and Davatzikos, C. (2008). Multiparametric tissue characterization of brain neoplasms and their recurrence using pattern classification of MR images. *Academic Radiology*, *15*(8), 966–977. doi: 10.1016/j.acra.2008.01.029

Wang, S., Kim, S., Chawla, S., Wolf, R. L., Knipp, D. E., Vossough, A., and Melhem, E. R. (2011). Differentiation between glioblastomas, solitary brain metastases, and primary cerebral lymphomas using diffusion tensor and dynamic susceptibility contrast-enhanced MR imaging. *American Journal of Neuroradiology*, *32*(3), 507–514. doi:10.3174/ajnr.a2333

Wang, S., Kim, S., Chawla, S., Wolf, R. L., Zhang, W. G., O'Rourke, D. M., Judy, K. D., Melhem, E. R., and Poptani, H. (2009). Differentiation between glioblastomas and solitary brain metastases using diffusion tensor imaging. *Neuroimaging*, *44*(3), 653–660. doi: 10.1016/j.neuroimage.2008.09.027

Wu, W., Chen, C., Chung, H., Gao, H., Hsueh, C., and Juan, C. (2002). Discrepant MR spectroscopic and perfusion imaging results in a case of malignant transformation of cerebral glioma. *AJNR American Journal of Neuroradiology*, *23*(10), 1775–1778.

Yamasaki, F., Arita, K., Hanaya, R., Hama, S., Ito, Y., Kurisu, K. et al. (2005). Apparent diffusion coefficient of human brain tumors at MR imaging. *Radiology*, *235*(3), 985–991.

Yang, S., Johnson, G., Law, M., Cha, S., Zagzag, D., Knopp, E., and Litt, A. (2003). Utility of Endothelial permeability measurements in differentiation between atypical and typical meningiomas using perfusion-weighted MRI. *Rivista di Neuroradiologia*, *16*(6), 1061–1062. doi:10.1177/197140090301600604

Yue, Q., Isobe, T., Shibata, Y., Anno, I., Kawamura, H., Yamamoto, Y., and Matsumura, A. (2008). New observations concerning the interpretation of magnetic resonance spectroscopy of meningioma. *European Radiology, 18*(12), 2901–2911. doi:10.1007/s00330-008-1079-6

Zacharaki, E. I., Kanas, V. G., and Davatzikos, C. (2011). Investigating machine learning techniques for MRI-based classification of brain neoplasms. *International Journal of Computer Assisted Radiology and Surgery, 6*(6), 821–828. doi:10.1007/s11548-011-0559-3

Zhang, H., Miao, J., Oudkerk, M., Rödiger, L. A., and Shen, T. (2008). Perfusion MR imaging for differentiation of benign and malignant meningiomas. *Neuroradiology, 50*, 525–530.

Zonari, P., Baraldi, P., and Crisi, G. (2007). Multimodal MRI in the characterization of glial neoplasms: The combined role of single-voxel MR spectroscopy, diffusion imaging and echo-planar perfusion imaging. *Neuroradiology, 49*(10), 795–803. doi:10.1007/s00234-007-0253-x

Index

Milton Keynes UK
Ingram Content Group UK Ltd.
UKHW050450071024
449327UK00014B/318

9 780367 876500